BIGEOMETRIC C.

A System with a Scale-Free Derivative

*

Michael Grossman
University of Lowell

1983

Archimedes Foundation
Box 240, Rockport
Massachusetts 01966

ISBN 0977117030

"for each successive class of phenomena,
a new calculus or a new geometry, as
the case might be, which might prove
not wholly inadequate to the subtlety
of nature."

Quoted, without citation,
by H. J. S. Smith; *Nature*,
Volume 8 (1873), page 450.

First Printing, 1983

P R E F A C E

The classical calculus, which was developed three hundred years ago by Newton, Leibniz, and their predecessors, has a derivative that is *origin-free*, i.e., it is invariant under all changes of origins in function arguments and values. The bigeometric calculus, which was created by Robert Katz and me in August 1970, has a derivative that is *scale-free*, i.e., it is invariant under all changes of scales (or units) in function arguments and values. That feature should appeal to scientists who seek ways to express laws in scale-free form.

The classical calculus and the bigeometric calculus are members of an infinite family of calculi that we constructed in the late sixties. All these calculi can be described simultaneously within the framework of a general theory. We decided to use the adjective "non-Newtonian" to indicate any of the calculi other than the classical calculus.

In 1971, we completed our *Non-Newtonian Calculus* [2], which is self-contained and includes a brief account of the bigeometric calculus, eight other specific non-Newtonian calculi, and the general theory.

As is the case with the classical and all the non-Newtonian calculi, the bigeometric calculus possesses the following:
> a distinctive method of measuring changes in function arguments;
> a distinctive method of measuring changes in function values;
> four operators: a gradient (i.e., an average rate of change), a derivative, a natural average, and an integral;
> a characteristic class of functions having a constant derivative;
> a Basic Theorem involving the gradient, derivative, and natural average;
> a Basic Problem whose solution motivates a simple definition of the integral in terms of the natural average;
> and two Fundamental Theorems which reveal that the derivative and integral are "inversely" related in an appropriate sense.

In the classical calculus, *differences* are used for measuring changes in arguments and in values. In the bigeometric calculus, *ratios* are used for measuring changes in arguments and in values.

The operators of the classical calculus are linear, but the operators of the bigeometric calculus are multiplicative.

When applied to specific functions the bigeometric operators yield numerical results that differ from those yielded by
the classical operators. And, of course, the bigeometric operators reflect different conceptions.

The classical derivative is constant for linear functions, but the bigeometric derivative is constant for power
functions.

Although there are many excellent ways of presenting the
principle ideas of the classical calculus, the novel presentation in Chapter 1 probably leads most naturally to the development of the bigeometric calculus in particular and to the
non-Newtonian calculi in general. Included in Chapter 1 is a
simple method of defining tangent lines without using limits
or derivatives.

Chapter 2 contains a development of the bigeometric calculus that parallels the development of the classical calculus
in Chapter 1.

Chapter 3 includes geometric arithmetic, which is a complete ordered field distinct from classical arithmetic (the
real number system), and which is used to reveal certain
structural similarities in the classical and bigeometric calculi. When applied to problems of change and accumulation,
classical arithmetic leads to the classical calculus; however,
it seems that no one had conceived the idea of using nonclassical arithmetics to construct new systems of calculus. Indeed, before 1967, apparently no one had used a nonclassical
arithmetic for any purpose, it having been long believed that
there is no distinctive value in the nonclassical arithmetics,
since they are all structurally equivalent to classical arithmetic. (Nonclassical arithmetic should be distinguished from
the nonstandard arithmetic developed by Abraham Robinson.)

Chapter 4 contains graphical interpretations of the bigeometric calculus; Chapter 5 includes a variety of heuristic
principles for selecting appropriate gradients, derivatives,
averages, and integrals; Chapter 6 includes a non-Cartesian
(analytic) geometry, which is a nonlinear model for plane Euclidean geometry.

Chapter 7 has discussions of bigeometric vectors, most of
which are curvilinear (unlike classical vectors which are all
rectilinear), and bigeometric centroids, which turn up again
in Chapter 8 on least-squares methods. The bigeometric method
of least squares provides what is probably the first suitable
rationale for the logarithmic-transformation technique that is
commonly used for fitting power curves to data. Chapter 9
contains, among other things, brief discussions of bigeometric
complex-numbers, weighted calculi, and meta-calculi.

Various digressions and comments have been placed in the NOTES at the ends of the sections. An annotated bibliography, a list of symbols, and an index have been provided at the end of the book.

Since this self-contained work is intended for a wide audience, including students, engineers, scientists, and mathematicians, many details that would not appear in a research report have been included, and proofs, most of which are straightforward, have been excluded. It is assumed, of course, that the reader has a working knowledge of the classical calculus.

For his invaluable assistance, I wish to thank Robert Katz, my friend, teacher, and colleague.

Suggestions and criticisms are invited.

Michael Grossman

University of Lowell
Department of Mathematics
Lowell, Massachusetts 01854

<u>C O N T E N T S</u>

PRELIMINARIES

In this book the word number means real number. The let-
ter R stands for the set of all numbers.

The arithmetic average of n numbers $v_1, \ldots, v_n$ is the number $(v_1 + \cdots + v_n) / n$.

If $r < s$, then the interval $[r,s]$ is the set of all numbers x such that $r \leq x \leq s$. (Only such intervals are used
here.) The interior of $[r,s]$ consists of all numbers x such
that $r < x < s$.

An arithmetic partition of an interval $[r,s]$ is any a-
rithmetic progression whose first term is r and last term is
s. An arithmetic partition that has exactly n terms is said
to be n-fold.

A point is any ordered pair of numbers, each of which is
called a coordinate of the point. A function is a set of
points, each distinct two of which have distinct first coordinates.

The arguments of a function are the first coordinates of
its points; the domain of a function is the set of all its arguments. A function is said to be on its domain and to be defined at each of its arguments.

The values of a function are the second coordinates of
its points; the range of a function is the set of all its values.

A discrete function is any function that has only a finite number of points.

1

If every two distinct points of a function f have distinct second coordinates, then f is <u>one-to-one</u> and its <u>in-
verse</u> is the one-to-one function consisting of all points
(y,x) for which (x,y) is a point of f.

The function exp is on R and assigns to each number x
the number e^x, where e is the base of the natural logarithm
function, ln. The function ln is the inverse of exp.

CHAPTER 1

The Classical Calculus

1.1 INTRODUCTION

> "The calculus was the first achievement of modern
> mathematics, and it is difficult to overestimate
> its importance. I think it defines more unequi-
> vocally than anything else the inception of mod-
> ern mathematics; and the system of mathematical
> analysis, which is its logical development, still
> constitutes the greatest technical advance in ex-
> act thinking."
>
> John von Neumann[1]

The distinguished mathematician was referring, of course,
to the <u>classical calculus</u>, the system of differential and integral calculus developed by Leibniz, Newton, and many predecessors, including Eudoxus, Archimedes, Kepler, Cavalieri,
Fermat, Wallis, and Barrow.[2]

In this chapter we present the basic ideas of the classical calculus in a manner that leads naturally to the construction of the bigeometric calculus in Chapter 2.

In the classical calculus, differences are used to measure changes (or deviations) in arguments and in values, and
sums are used to accumulate (or combine) arguments and to accumulate values. It is fitting, therefore, that Leibniz often
referred to the classical calculus as a "calculus of differences and sums."[3] Indeed, Carl B. Boyer pointed out that "...
Leibniz looked upon the operation of finding 'differences' as
fundamental in his...calculus."[4]

N O T E S

1. Von Neumann's famous remark appeared originally in *The Works of the Mind* (Chicago: University of Chicago Press, 1947), an anthology of interesting articles edited by R. B. Heywood.

2. Although the term "classical analysis" is often used, the term "classical calculus" has appeared rarely, presumably because there was only one calculus prior to the discovery of the non-Newtonian calculi. The term was used, for example, by Stanley L. Jaki in *The Relevance of Physics* (The University of Chicago Press, 1966), an absorbing book on the history of physics.

The adjective "non-Newtonian" may have been first used in 1909 by G. N. Lewis and R. C. Tolman, who were referring to the theory of relativity.

3. In 1696, Leibniz wrote: "I have brought matters so far with my infinitesimal calculus of differences and sums that many problems can now be solved in mathematical physics which one could not even venture to try before." This appeared in *Philosophical Papers and Letters of Gottfried Wilhelm Leibniz*, Vol. II, ed. and trans. L. E. Loemker (Chicago: University of Chicago Press, 1956), p.768.

4. Carl B. Boyer, *The History of the Calculus and Its Conceptual Development* (New York: Dover reprint, 1949), p.206.

1.2 LINEAR FUNCTIONS

> "...the fundamental idea of Calculus, namely the 'local' approximation of functions by *linear* functions."
>
> Jean Dieudonné[1]

In the classical calculus, the linear functions are the standards to which other functions are compared.

A <u>linear function</u> is any function ℓ on R such that $\ell(x) = mx + c$, where m and c are constants.[2]

Each linear function ℓ is uniform in the following sense:

For any intervals $[r_1, s_1]$ and $[r_2, s_2]$,

if $s_1 - r_1 = s_2 - r_2$,

then $\ell(s_1) - \ell(r_1) = \ell(s_2) - \ell(r_2)$;

that is, equal differences in arguments yield equal

differences in values.[3]

In particular, the number $\ell(b) - \ell(a)$ is the same for any two

numbers a and b such that $b - a = 1$, a fact that suggests the

next definition.

N O T E S

1.   Jean Dieudonné, *Foundations of Modern Analysis* (New York: Academic Press, 1960), p. 141.   In regard to the classical calculus of functions of one variable, we prefer Whitehead's remark in Section 1.4.

2.   In Section 1.4 we shall use the fact that there is precisely one linear function containing any two given points with different first coordinates.

3.   The linear functions are the *only* continuous functions on R such that equal differences in arguments yield equal differences in values.

1.3   CLASSICAL SLOPE

The <u>classical slope</u> of a linear function ℓ is the number

$\ell(b) - \ell(a)$, where a and b are any two numbers such that $b - a$

$= 1$.

Though that definition is seldom used, it is well-known,

is extremely simple, and is suggestive of other kinds of

slopes, one of which is defined in Section 2.3.[1]

Of course the classical slope of the linear function $\ell(x)$

= mx + c turns out to be m.

N O T E

1. Though this definition of classical slope was used, in effect, by Karl Pearson (*The Grammar of Science*) and undoubtedly
by many others, most, if not all, textbook writers use the
"rise over run" definition, which, though not incorrect, fails
to suggest the extensions that are required for the non-Newtonian calculi. Indeed, one of our early difficulties arose
from our failure to use the simpler definition of classical
slope.

1.4 THE CLASSICAL GRADIENT

> "The importance of the differential calculus arises
> from the very nature of the subject, which is the
> systematic consideration of the [gradients] of func-
> tions."
>
> Alfred North Whitehead[1]

The differential branch of the classical calculus is
rooted in the concept of the average rate of change, which we
prefer to call the classical gradient.[2]

The classical gradient of a function f on an interval
[r,s] is denoted by $G_r^s f$ and is defined to be the classical
slope of the linear function containing the points (r,f(r))
and (s,f(s)).

It turns out that

$$G_r^s f = \frac{f(s) - f(r)}{s - r} \, ,$$

thus providing a simple way of calculating $G_r^s f$. Nevertheless,
the reader is urged to conceive $G_r^s f$ as the classical slope of

the linear function containing $(r,f(r))$ and $(s,f(s))$, since that conception lends itself to useful generalizations, one of which appears in Section 2.4.[3]

The operator G is

<u>Additive</u>: $G_r^s(f + g) = G_r^s f + G_r^s g,$

<u>Subtractive</u>: $G_r^s(f - g) = G_r^s f - G_r^s g,$

<u>Homogeneous</u>: $G_r^s(c \cdot f) = c \cdot G_r^s f,$ c any constant.

Though those three facts can be expressed in one equation, we prefer to list them separately.

Of course, the classical gradient of a linear function on any interval is equal to its classical slope.

If each point (x,y) of a function is changed to $(x + a, y + b)$, where a and b are constants, then a <u>change of origins</u> has been made in the function. It turns out that the classical gradient is <u>origin-free</u>, that is, invariant under every change of origins.

N O T E S

1. A. N. Whitehead, *An Introduction to Mathematics* (New York: Henry Holt and Company, 1911).

2. The term "gradient" is better suited for our purposes since it can be readily modified by appropriate adjectives. Of course, "gradient" is also used in vector analysis, but that subject does not concern us here.

3. Infinitely-many gradients were first defined in [2].

1.5 THE CLASSICAL DERIVATIVE

> "It was Fermat, primarily, who introduced the modern
> idea of the tangent to a curve at a given point P.
> In essence, he took a second point Q on the curve,
> found the slope of the secant line PQ, and from this,
> by permitting Q to tend toward coincidence with P,
> he calculated the slope of the tangent. This method
> rightly earned him the title of inventor of the dif-
> ferential calculus."
>
> Boyer & Neugebauer[1]

In this section f is assumed to be a function defined at

least on an interval containing the number a in its interior.

If the following limit[2] exists, we denote it by [Df](a), call

it the <u>classical derivative of f at a</u>, and say that f is clas-

sically differentiable at a:

$$\lim_{x \to a} \frac{f(x) - f(a)}{x - a} \ .$$

The <u>classical derivative of f</u>, denoted by Df, is the

function that assigns to each number t the number [Df](t), if

it exists.[3]

The operator D is additive, subtractive, and homogeneous;

that is, if [Df](a) and [Dg](a) exist, then

$$[D(f + g)](a) = [Df](a) + [Dg](a),$$

$$[D(f - g)](a) = [Df](a) - [Dg](a),$$

$$[D(c \cdot f)](a) = c \cdot [Df](a), \quad c \text{ any constant.}$$

Furthermore, the operator D is origin-free.

If ℓ is a linear function, then $D\ell$ has a constant value

equal to the classical slope of ℓ. Indeed, only linear func-

tions have classical derivatives that are constant on R. In

particular, if ℓ is a constant function on R, then $D\ell$ is everywhere equal to 0.

Next, the familiar concept of tangent line is defined in a simple way without using limits or derivatives. This definition has the virtue that it can be generalized in an important way.[4]

The tangent to f at the point (a,f(a)) is the unique linear function g, if it exists, that possesses the following two properties.

1) The linear function g contains (a,f(a)).

2) For each linear function ℓ containing (a,f(a)) and distinct from g, there is a positive number p such that for every number x in [a - p, a + p] but distinct from a,

$$|g(x) - f(x)| < |\ell(x) - f(x)|.$$

(Roughly, the tangent is locally closer to f than any other linear function.)

It can be proved that [Df](a) exists if and only if f has a tangent at (a,f(a)); and if [Df](a) does exist, it equals the classical slope of that tangent.[5]

Two functions are tangent at a common point if and only if they have the same tangent there.

N O T E S

1. The remark is from the article "History of Mathematics," which appears in some editions of the *Encyclopaedia Britannica*, e.g., in the 1970 edition.

2. We do not discuss limits and continuity in this book, since these concepts are identical in the classical and bigeo-

metric calculi. However, for most other non-Newtonian calculi one must introduce different limit and continuity concepts.

3. Though Laplace rightly remarked that "Leibniz has enriched the differential calculus by a very happy notation," the Leibniz "d" notation is not convenient for our purposes.

4. See page 41 of [2].

5. Robert Katz and I formulated this definition of the tangency concept. We are grateful to Charles Rockland for furnishing a proof of the equivalence of our definition to the usual one.

1.6 THE ARITHMETIC AVERAGE

> "...averaging processes have proved themselves power-ful tools in analysis."
>
> Einar Hille

The concept of arithmetic average of a continuous function on an interval plays an important role in our treatment of the classical calculus.

The arithmetic average of a continuous function f on an interval $[r,s]$ is denoted by $A_r^s f$ and is defined to be the limit of the convergent sequence whose nth term is the arithmetic average of $f(a_1), \ldots, f(a_n)$, where $a_1, \ldots, a_n$ is the n-fold arithmetic partition of $[r,s]$.

The operator A is additive, subtractive, and homogeneous; that is, if f and g are continuous on $[r,s]$, then

$$A_r^s(f + g) = A_r^s f + A_r^s g,$$
$$A_r^s(f - g) = A_r^s f - A_r^s g,$$
$$A_r^s(c \cdot f) = c \cdot A_r^s f, \quad c \text{ any constant.}$$

The operator A is characterized by the following three properties. (This use of the term "characterized" indicates that no other operator possesses all three properties.)

For any interval [r,s] and any constant function k(x) = c on [r,s],

$$A_r^s k = c.$$

For any interval [r,s] and any continuous functions f and g on [r,s], if $f(x) \leq g(x)$ for every number x in [r,s], then

$$A_r^s f \leq A_r^s g.$$

For any numbers r, s, t such that r < s < t, and any continuous function f on [r,t],

$$(s - r) \cdot A_r^s f + (t - s) \cdot A_s^t f = (t - r) \cdot A_r^t f.$$

It is an interesting fact that the arithmetic average of a linear function on [r,s] is equal to its value at the arithmetic average of r and s, which is equal to the arithmetic average of its values at r and s.

1.7 THE BASIC THEOREM OF CLASSICAL CALCULUS

"No great theory lacks a kernel."
Anonymous

For many years we have been guided by the idea that the "kernel" of the classical calculus is neither the First nor the Second Fundamental Theorem of Classical Calculus, but rather a well-known result that we call the Basic Theorem of

Classical Calculus. Let us begin with its discrete analogue, which is a proposition that concerns discrete functions and appropriately conveys the spirit of the theorem.

The Discrete Analogue of the

Basic Theorem of Classical Calculus

If h is a discrete function whose arguments $a_1, \ldots,$ a_n constitute an arithmetic partition of $[r,s]$, then the arithmetic average of the classical gradients of h on the intervals $[a_{i-1}, a_i]$, $i = 2, \ldots, n$, is equal to the classical gradient of h on $[r,s]$.

The foregoing result suggests the following important theorem.[1]

The Basic Theorem of Classical Calculus

If Dh is continuous on $[r,s]$, then its arithmetic average on $[r,s]$ equals the classical gradient of h on $[r,s]$, that is,

$$A_r^s(Dh) = \frac{h(s) - h(r)}{s - r}.$$

In view of this theorem we say that the arithmetic average fits naturally into the scheme of classical calculus.

N O T E

1. The Basic Theorem of Classical Calculus is a well-known variant of the Second Fundamental Theorem of Classical Calculus, which is stated in Section 1.10. Every non-Newtonian calculus has its own Basic Theorem that is an extension of a simple algebraic identity and has the following form, in which the underlined words have a special meaning depending on the nature of the particular calculus:

If the underline{derivative} of a function is underline{continuous} on an interval, then its underline{average} thereon equals the underline{gradient} of the function thereon. This is discussed fully in [2].

One attractive feature of the Basic Theorem is the existence of its discrete analogue, which can be understood before the introduction of limits, derivatives, and integrals.

1.8 THE BASIC PROBLEM OF CLASSICAL CALCULUS

Suppose that the value of a function h is known at an argument r, and suppose that f, the classical derivative of h, is continuous and known at each number in [r,s]. Find h(s).

Solution

By the Basic Theorem of Classical Calculus,

$$A_r^s f = A_r^s(Dh) = \frac{h(s) - h(r)}{s - r} \, .$$

Hence,

$$h(s) = h(r) + (s - r) \cdot A_r^s f.$$

The number $(s - r) \cdot A_r^s f$ in the foregoing solution arises with sufficient frequency to warrant a special name, the classical integral of f on [r,s], which is introduced in the next section.

Thus, the Basic Theorem of Classical Calculus, which involves the arithmetic average, classical derivative, and classical gradient, provides for the Basic Problem of Classical Calculus an immediate solution, which, in turn, motivates our definition of the classical integral in terms of the arithmetic average.[1]

N O T E

1. Ideally, the introduction of a new concept should be pre-
ceded by a problem whose solution suggests the concept. However, that cannot always be done in a natural way.

1.9 THE CLASSICAL INTEGRAL

> "The theory of integration is concerned with finding
> averages of functions."
>
> Edwin Hewitt[1]

The <u>classical integral</u> of a continuous function f on an

interval [r,s] is denoted by $\int_r^s f$ and is defined to be the num-

ber $(s - r) \cdot A_r^s f$. We set $\int_r^r f = 0$.

It turns out that $\int_r^s f$ equals the limit of the convergent

sequence whose nth term is the sum

$$k_n \cdot f(a_1) + \cdots + k_n \cdot f(a_{n-1}),$$

where $a_1, \ldots, a_n$ is the n-fold arithmetic partition of [r,s],

and k_n is the common value of

$$a_2 - a_1, \; a_3 - a_2, \; \ldots, \; a_n - a_{n-1}.$$

The operator $\int$ is additive, subtractive, and homogeneous;

that is, if f and g are continuous on [r,s], then

$$\int_r^s (f + g) = \int_r^s f + \int_r^s g,$$

$$\int_r^s (f - g) = \int_r^s f - \int_r^s g,$$

$$\int_r^s (c \cdot f) = c \cdot \int_r^s f, \quad c \text{ any constant.}$$

The operator $\int$ is characterized by the following three properties.

For any interval [r,s] and any constant function k(x) = c on [r,s],

$$\int_r^s k = (s - r) \cdot c.$$

For any interval [r,s] and any continuous functions f and g on [r,s], if $f(x) \leq g(x)$ for every number x in [r,s], then

$$\int_r^s f \leq \int_r^s g.$$

For any numbers r, s, t such that r < s < t, and any continuous function f on [r,t],

$$\int_r^s f + \int_s^t f = \int_r^t f.$$

In Section 5.3 there is an important application of the foregoing characterization.

The classical integral can be characterized in other interesting and useful ways.[2]

N O T E S

1. The remark occurs on page 50 of Hewitt's preliminary edition of *Theory of Functions of a Real Variable* (New York: Holt, Rinehart, and Winston, 1960). In their *Real and Abstract Analysis* (New York: Springer-Verlag Inc., 1965), Hewitt and Stromberg asserted that "Integration from one point of view is an averaging process for functions...."

It is surprising, nevertheless, that averages do not seem to play a central role in modern analysis. For example, only rarely does one encounter the geometric, quadratic, and harmonic averages in integration theory. (These averages are discussed in [2], [3], [5], and [7].)

In a special course he gave at Harvard University in 1958, Robert Katz defined the classical integral in terms of the arithmetic average, believing that such a procedure is intuitively more satisfying. Subsequently, in his courses at Tufts University, he always used that technique, which found considerable favor with the students and the engineering faculty, especially since the use of the arithmetic average simplifies the presentation of many scientific concepts. (For example, see Section 5.4.)

2. An excellent characterization of the classical integral may be found in Chapters 5 and 6 of *Calculus* (New York: W. W. Norton, 1973) by Leonard Gillman and Robert H. McDowell.

1.10 THE FUNDAMENTAL THEOREMS OF CLASSICAL CALCULUS

> "Areas by integration had been found, through summations, by earlier mathematicians from Archimedes to Wallis; and differentiations had been carried out by Fermat. It remained for Newton and Gottfried Leibniz to discover the fundamental principle of the calculus - that integrations can be performed far more easily by inverting the process of differentiation."
>
> Boyer & Neugebauer[1]

The classical derivative and classical integral are "inversely" related in the sense indicated by the following two theorems.

First Fundamental Theorem of Classical Calculus

If f is continuous on [r,s], and
$$g(x) = \int_r^x f, \text{ for every number } x \text{ in } [r,s],$$
then
$$Dg = f, \text{ on } [r,s].[2]$$

Second Fundamental Theorem of Classical Calculus

If Dh is continuous on [r,s], then
$$\int_r^s (Dh) = h(s) - h(r).$$

The latter theorem is a simple consequence of the Basic Theorem of Classical Calculus.

This concludes our brief discussion of the classical calculus.

N O T E S

1. For a citation see Note 1 to Section 1.5.

2. The following theorem reveals another interesting property
of the function $\int_r^x f$.

Let $r < a < s$ and assume f is continuous on $[r,s]$. Then

$$G_a^s \left[\int_r^x f \right] = A_a^s f.$$

CHAPTER 2

The Bigeometric Calculus

2.1 INTRODUCTION

> "The language of analysis...and its notations...are
> so many germs of new calculi."
>
> Laplace

In the bigeometric calculus, ratios are used to measure changes (or deviations) in arguments and in values, and products are used to accumulate (or combine) arguments and to accumulate values. Indeed, ratios and products play the same role in the bigeometric calculus that differences and sums play in the classical calculus.

Furthermore, the operators of the bigeometric calculus are applied only to functions having positive arguments and values.[1]

It is certainly not unusual to measure deviations by ratios rather than by differences. For instance, during the Renaissance many scholars, including Galileo, discussed the following problem:

> Two estimates, 10 and 1000, are proposed as the value of a horse. Which estimate, if any, deviates more from the true value of 100?

The scholars who maintained that the deviations should be measured by differences concluded that the estimate of 10 was closer to the true value. However, Galileo eventually maintained that the deviations should be measured by ratios and that accordingly the two estimates deviated equally from the true value.

Four hundred years later the controversy about ratios and differences was still alive, as indicated by the following remark by William H. Kruskal.[2]

> "In recent discussions of possible relationships between cigarette smoking and lung cancer, controversy arose over whether ratios or differences of mortality rates were of central importance. The choice may lead to quite different conclusions."

It is also worth noting that the ear's subjective comparison of loudness is apparently achieved by sensing ratios, not differences, of sound levels.[3]

Now let us consider the following problem. At the beginning of 1981, the wage-rate (in dollars per hour) at a certain company was w_0 and the cost-of-living index for the United States was c_0. At the end of 1981, the amounts were w_1 and c_1, respectively. Company and union negotiators had agreed at the beginning of 1981 that thereafter the wage-rate would be adjusted to reflect changes in the cost-of-living index. Assuming that the cost-of-living index is always increasing and that the wage-rate changes "uniformly" and continuously relative to the cost-of-living index, find the wage-rate w_t at time t in terms of the constants c_0, c_1, w_0, w_1 and the cost-of-living index c_t at time t. There is no unique solution to this problem; we shall give two reasonable solutions.

Solution 1. Since the wage-rate changes "uniformly" relative to the cost-of-living index, we may reasonably assume that

equal differences in the cost-of-living index give rise to e-
qual differences in the wage-rate. Furthermore, since the
wage-rate changes continuously relative to the cost-of-living
index, it can be proved that

$$w_t = w_0 + \left[\frac{w_1 - w_0}{c_1 - c_0} \right] (c_t - c_0).$$

Notice that the expression within the brackets represents a
classical gradient.

Solution 2. Since the wage-rate changes "uniformly" relative
to the cost-of-living index, we may reasonably assume that e-
qual ratio changes in the cost-of-living index give rise to
equal ratio changes in the wage-rate. Furthermore, since the
wage-rate changes continuously relative to the cost-of-living
index, it can be proved that

$$w_t = w_0 \left[\left(\frac{w_1}{w_0} \right)^{1/(\ln c_1 - \ln c_0)} \right]^{\ln c_t - \ln c_0}.$$

In Section 2.4, we shall see that the expression within the
brackets represents a new gradient that plays a fundamental
role in the bigeometric calculus.

The operators of the bigeometric calculus will be inter-
pretated graphically in Chapter 4, and heuristic principles
for their application will be given in Chapter 5.

The following notation and terminology will be useful.

The symbol R_+ stands for the set of all positive numbers.

A positive interval is any interval [r,s] such that 0 <
r < s.

A <u>bipositive point</u> is any point both of whose coordinates are positive. A <u>bipositive function</u> is any function all of whose points are bipositive.

Finally, it will be convenient to use the prefix "*-" to stand for "bigeometric" or "bigeometrically," whichever is grammatically appropriate.

N O T E S

1. In Section 6.10 of [2], bigeometric-type calculi applicable to certain other functions are discussed.

2. The remark was made in Kruskal's article "Statistics: The Field," which appears in Volume 15 of the *International Encyclopedia of the Social Sciences*, edited by David L. Sills and published by The Macmillan Company and The Free Press in 1968.

3. See, e.g., William C. Vergara, *Science, the Never-Ending Quest* (Harper & Row, 1965), p. 132.

2.2 POWER FUNCTIONS

> "It is only by the hook of the analogy...that science
> succeeds in extending its domain."
>
> Loren Eiseley

We know that each linear function is uniform in the sense that equal differences in its arguments yield equal differences in its values. This suggests that we seek a class of functions having an analogous property with respect to ratios of arguments and ratios of values.

By a <u>power function</u> we mean any bipositive function p on R_+ such that $p(x) = cx^m$, where c and m are constants and c >

0. Clearly, every constant bipositive function on R_+ is a
power function.[1]

Each power function p is uniform in the following sense:

For any positive intervals $[r_1,s_1]$ and $[r_2,s_2]$,

if $s_1/r_1 = s_2/r_2$, then $p(s_1)/p(r_1) = p(s_2)/p(r_2)$;

that is, equal ratios of arguments yield

equal ratios of values.[2]

In particular, the number $p(b)/p(a)$ is the same for any two
positive numbers a and b such that $b/a = e$, a fact that suggests the definition in the next section.

The power functions play the same role in the *-calculus
that the linear functions play in the classical calculus. Indeed, in the *-calculus the power functions are the standards
to which other functions are compared.

N O T E S

1. In Section 2.4 we shall use the fact that there is precisely one power function containing any two given bipositive
points with different first coordinates.

2. The power functions are the *only* continuous bipositive
functions on R_+ such that equal ratios of arguments yield e-
qual ratios of values.

2.3 BIGEOMETRIC SLOPE

"When we study things that vary, it is only that we
may find what is uniform and constant there."
Louis Poinsot[1]

We defined the classical slope of a linear function ℓ to be the number $\ell(b) - \ell(a)$, where a and b are any two numbers such that b - a = 1.

Similarly, the *-slope of a power function p is the positive number $p(b)/p(a)$, where a and b are any two positive numbers such that b/a = e. (Our reason for using e rather than some other positive constant is discussed in Section 3.3.)

The *-slope of the power function $p(x) = cx^m$ turns out to be e^m.

N O T E

1. Louis Poinsot, *Elemens de Statique*.

2.4 THE BIGEOMETRIC GRADIENT

> "To new concepts correspond, necessarily, new symbols."
>
> Gauss

The classical gradient of a function f on an interval [r,s] was defined to be the classical slope of the linear function containing the points (r,f(r)) and (s,f(s)). This suggests the following definition.

The *-gradient of a bipositive function f on a positive interval [r,s] is denoted by $\overset{*}{G}{}^s_r f$ and is defined to be the *-slope of the power function containing the bipositive points (r,f(r)) and (s,f(s)).

It turns out that

$$\overset{*s}{G_r}f = \left[\frac{f(s)}{f(r)} \right]^{\frac{1}{\ln s - \ln r}} ,$$

thus providing a convenient formula for $\overset{*s}{G_r}f$.

The operator $\overset{*}{G}$ is

    <u>Multiplicative</u>: $\overset{*s}{G_r}(f \cdot g) = \overset{*s}{G_r}f \cdot \overset{*s}{G_r}g$,

    <u>Divisional</u>: $\overset{*s}{G_r}(f / g) = \overset{*s}{G_r}f / \overset{*s}{G_r}g$,

    <u>Involutional</u>: $\overset{*s}{G_r}(f^c) = (\overset{*s}{G_r}f)^c$, c any constant.

Of course, the *-gradient of a power function on any positive interval is equal to its *-slope.

If each point (x,y) of a bipositive function is changed to (px, qy), where p and q are positive constants, then a <u>change of scales</u> (or units) has been made in the function. It turns out that the *-gradient is <u>scale-free</u>, that is, invariant under every change of scales.

When r = s, the expression for the *-gradient yields the indeterminate form 1^∞, in contrast to the indeterminate form 0/0 yielded by the expression for the classical gradient.

2.5 THE BIGEOMETRIC DERIVATIVE

In this section f is assumed to be a bipositive function defined at least on a positive interval containing the positive number a in its interior. If the following limit exists and is positive,[1] we denote it by $[\overset{*}{D}f](a)$, call it the <u>*-derivative of f at a</u>, and say that f is *-differentiable at a:

$$\lim_{x \to a} \left[\frac{f(x)}{f(a)} \right]^{1/(\ln x - \ln a)}$$

It can be proved that $[\overset{*}{D}f](a)$ and $[Df](a)$ coexist; that is, if either exists then so does the other. Moreover, if they do exist, then

$$[\overset{*}{D}f](a) = \exp \left\{ a \cdot [Df](a) / f(a) \right\},$$

and $[\overset{*}{D}f](a)$ equals the *-slope of the unique power function that is tangent to f at $(a,f(a))$.

The *-derivative of f, denoted by $\overset{*}{D}f$, is the function that assigns to each positive number t the number $[\overset{*}{D}f](t)$, if it exists.

If p is a power function, then $\overset{*}{D}p$ has a constant value equal to the *-slope of p. Indeed, only power functions have *-derivatives that are constant on R_+. In particular, if p is a constant bipositive function on R_+, then $\overset{*}{D}p = 1$ on R_+.[2]

The operator $\overset{*}{D}$ is multiplicative, divisional, and involutional; that is, if $[\overset{*}{D}f](a)$ and $[\overset{*}{D}g](a)$ exist, then

$$[\overset{*}{D}(f \cdot g)](a) = [\overset{*}{D}f](a) \cdot [\overset{*}{D}g](a),$$

$$[\overset{*}{D}(f / g)](a) = [\overset{*}{D}f](a) / [\overset{*}{D}g](a),$$

$$[\overset{*}{D}f^c](a) = ([\overset{*}{D}f](a))^c, \quad c \text{ any constant.}$$

It is worth noting that if $w(x) = \exp x$ on R_+, then $\overset{*}{D}w = w$, that is, w equals its *-derivative.

The operator $\overset{*}{D}$ is scale-free. Hence, $\overset{*}{D}$ is independent of the scales (or units) used for function arguments and values, a fact that might interest scientists who wish to express laws in scale-free form.

We noted above that

$$[\overset{*}{D}f](a) \;=\; \exp\left\{ a \cdot [Df](a) \,/\, f(a) \right\} .$$

Since economists call the expression within the braces the e-
lasticity of f at a, we refer to $[\overset{*}{D}f](a)$ as the resiliency of
f at a. Thus, elasticity is the natural logarithm of resiliency.

We believe that resiliency will prove to be more useful
than elasticity because the former is the derivative in a complete system of calculus (the naturalness of which is exhibit-
ed in this chapter), whereas it appears to be impossible to
construct a complete, natural system of calculus in which the
derivative is the elasticity.

Perhaps the psychophysicists will find some use for the
*-calculus, since one of their basic laws may be stated thus:
The resiliency of the stimulus-sensation function is constant.
(That constant is determined by the nature of the stimulus.)

The *-calculus may also prove to be useful in biology,
for a fundamental law of growth is the following: If f is the
function relating the size of one organ to the size of any
other given organ in the same body at the same instant, then,
within certain time limits, the resiliency of f is constant.

N O T E S

1. It is possible for the limit to be 0. Our reason for re-
quiring the limit to be positive is best explained in the context of the general theory of the non-Newtonian calculi and is
given in Section 6.10 of [2].

2. Therefore, for example, if $p(x) = 1$ on R_+, then

$[\overset{*}{D}(p + p)](3) \neq [\overset{*}{D}p](3) + [\overset{*}{D}p](3)$. Thus, the operator $\overset{*}{D}$ is not additive.

2.6 THE BIGEOMETRIC AVERAGE

The arithmetic average of a continuous function f on an interval [r,s] was defined to be the limit of the convergent sequence whose nth term is the arithmetic average of $f(a_1)$, ...,$f(a_n)$, where $a_1,...,a_n$ is the n-fold arithmetic partition of [r,s].

Our definition of the *-average of a continuous biposi- tive function on a positive interval is based, not on arithme- tic averages and arithmetic partitions, but on geometric aver- ages and geometric partitions, which are defined next.[1]

The geometric average of n positive numbers $w_1,...,w_n$ is the positive number $(w_1 w_2 \cdots w_n)^{1/n}$.

A geometric partition of a positive interval [r,s] is any geometric progression whose first term is r and last term is s. A geometric partition that has exactly n terms is said to be n-fold.

The *-average of a continuous bipositive function f on a positive interval [r,s] is denoted by $\overset{*}{A}{}_r^s f$ and is defined to be the positive limit of the convergent sequence whose nth term is the geometric average of $f(a_1),...,f(a_n)$, where $a_1,...,a_n$ is the n-fold geometric partition of [r,s].[2]

The operator $\overset{*}{A}$ is multiplicative, divisional, and involu-

tional; that is, if f and g are continuous bipositive functions on a positive interval [r,s], then

$$\overset{*s}{\underset{r}{A}}(f \cdot g) = \overset{*s}{\underset{r}{A}}f \cdot \overset{*s}{\underset{r}{A}}g,$$

$$\overset{*s}{\underset{r}{A}}(f / g) = \overset{*s}{\underset{r}{A}}f / \overset{*s}{\underset{r}{A}}g,$$

$$\overset{*s}{\underset{r}{A}}(f^c) = (\overset{*s}{\underset{r}{A}}f)^c, \quad c \text{ any constant.}$$

The operator $\overset{*}{A}$ is characterized by the following three
properties.

For any positive interval [r,s] and any constant bipositive function k(x) = c on [r,s],

$$\overset{*s}{\underset{r}{A}}k = c.$$

For any positive interval [r,s] and any continuous
bipositive functions f and g on [r,s], if f(x) $\leq$
g(x) for every number x in [r,s], then

$$\overset{*s}{\underset{r}{A}}f \leq \overset{*s}{\underset{r}{A}}g.$$

For any positive numbers r, s, t such that r < s < t,
and any continuous bipositive function f on [r,t],

$$[\overset{*s}{\underset{r}{A}}f]^{\ln s - \ln r} \cdot [\overset{*t}{\underset{s}{A}}f]^{\ln t - \ln s}$$
$$= [\overset{*t}{\underset{r}{A}}f]^{\ln t - \ln r}.$$

It is an interesting fact that the *-average of a power
function on a positive interval [r,s] is equal to its value
at the geometric average of r and s, which is equal to the
geometric average of its values at r and s.

N O T E S

1. By using geometric averages and arithmetic partitions, one
can obtain the well-known geometric average of f on [r,s],
which, for example, is discussed in [3]. By using arithmetic
averages and geometric partitions, one can obtain the anageo-

metric average of f on [r,s], which was first defined in [2].

2. Although infinitely-many averages have long been known, it appears that the *-average was unknown prior to our discovery of it in August of 1970. The *-average is discussed in [2] and briefly in [7].

2.7 THE BASIC THEOREM OF BIGEOMETRIC CALCULUS

As in the discussion of the Basic Theorem of Classical Calculus (Section 1.7), we begin with a discrete analogue.

The Discrete Analogue of the

Basic Theorem of *-Calculus

If h is a discrete bipositive function whose arguments $a_1, \ldots, a_n$ constitute a geometric partition of
a positive interval [r,s], then the geometric average of the *-gradients of h on the intervals $[a_{i-1},$
$a_i]$, i = 2,...,n, is equal to the *-gradient of h on
[r,s].

The foregoing result suggests the following important theorem.

The Basic Theorem of *-Calculus

If $\overset{*}{D}h$ is continuous on a positive interval [r,s],
then its *-average on [r,s] equals the *-gradient of
h on [r,s], that is,

$$\overset{*s}{\underset{r}{A}}(\overset{*}{D}h) = \left[\frac{h(s)}{h(r)}\right]^{1/(\ln s - \ln r)}$$

In view of this theorem we say that the *-average fits

naturally into the scheme of *-calculus.[1]

N O T E

1. As explained fully in [2], each of the various non-Newtonian calculi possesses an average (of functions) that fits naturally therein. Furthermore, in [7], it is shown how such averages can be used to construct an infinite family of means of
two positive numbers.

2.8 THE BASIC PROBLEM OF BIGEOMETRIC CALCULUS

Suppose that the value of a bipositive function h is
known at an argument r, and suppose that f, the *-
derivative of h, is continuous and known at each
number in [r,s]. Find h(s).

Solution

By the Basic Theorem of *-Calculus,

$$\overset{*s}{\underset{r}{A}}f = \overset{*s}{\underset{r}{A}}(\overset{*}{D}h) = \left[\frac{h(s)}{h(r)} \right]^{1/(\ln s - \ln r)} .$$

Hence,

$$h(s) = h(r) \cdot (\overset{*s}{\underset{r}{A}}f)^{\ln s - \ln r} .$$

The number $(\overset{*s}{\underset{r}{A}}f)^{\ln s - \ln r}$ in the foregoing solution will
arise with sufficient frequency to warrant a special name, the
*-integral of f on [r,s], which is introduced in the next
section.

Thus, the Basic Theorem of *-Calculus, which involves the
*-average, *-derivative, and *-gradient, provides for the Ba-
sic Problem of *-Calculus an immediate solution, which, in

turn, motivates our definition of the *-integral in terms of the *-average.[1]

N O T E

1. Every non-Newtonian calculus has a Basic Problem, whose solution via its Basic Theorem motivates a definition of the integral in that calculus.

2.9 THE BIGEOMETRIC INTEGRAL

> "It seems to be expected of every pilgrim up the slopes of the mathematical Parnassus, that he will at some point or other of his journey sit down and invent a definite integral or two towards the increase of the common stock."
>
> J. J. Sylvester[1]

The *-integral of a continuous bipositive function f on a positive interval [r,s] is denoted by $\int_{r}^{*s} f$ and is defined to be the positive number $[\overset{*s}{\underset{r}{A}} f]^{\ln s - \ln r}$. We set $\int_{r}^{*r} f = 1$.

It turns out that $\int_{r}^{*s} f$ equals the positive limit of the convergent sequence whose nth term is the product

$$[f(a_1)]^{\ln k_n} \cdot [f(a_2)]^{\ln k_n} \cdots [f(a_{n-1})]^{\ln k_n},$$

where $a_1, \ldots, a_n$ is the n-fold geometric partition of [r,s], and k_n is the common value of

$$a_2/a_1, \ a_3/a_2, \ \ldots, \ a_n/a_{n-1}.$$

The operator $\int^{*}$ is multiplicative, divisional, and invo-

lutional; that is, if f and g are continuous bipositive functions on a positive interval [r,s], then

$$\overset{*}{\int_r^s}(f \cdot g) = \overset{*}{\int_r^s}f \cdot \overset{*}{\int_r^s}g,$$

$$\overset{*}{\int_r^s}(f / g) = \left[\overset{*}{\int_r^s}f\right] / \left[\overset{*}{\int_r^s}g\right],$$

$$\overset{*}{\int_r^s}(f^c) = \left[\overset{*}{\int_r^s}f\right]^c, \quad \text{c any constant.}$$

The operator $\overset{*}{\int}$ is characterized by the following three
properties.

For any positive interval [r,s] and any constant bipositive function k(x) = c on [r,s],

$$\overset{*}{\int_r^s}k = c^{\ln s - \ln r}.$$

For any positive interval [r,s] and any continuous
bipositive functions f and g on [r,s], if f(x) ≤
g(x) for every number x in [r,s], then

$$\overset{*}{\int_r^s}f \le \overset{*}{\int_r^s}g.$$

For any positive numbers r, s, t such that r < s < t,
and any continuous bipositive function f on [r,t],

$$\overset{*}{\int_r^s}f \cdot \overset{*}{\int_s^t}f = \overset{*}{\int_r^t}f.$$

In Section 5.3, there is an application of the foregoing
characterization.

N O T E

1. Sylvester's remark appears on page 328 of Robert E. Moritz's book *On Mathematics And Mathematicians* (New York: Dover Publications Inc., 1958), which was originally entitled *Memorabilia Mathematica*.

2.10 THE FUNDAMENTAL THEOREMS OF BIGEOMETRIC CALCULUS

The *-derivative and *-integral are "inversely" related in the sense indicated by the following two theorems, the second of which is a simple consequence of the Basic Theorem of *-Calculus.

First Fundamental Theorem of *-Calculus

If f is a continuous bipositive function on a positive interval [r,s], and

$$g(x) = \int_{r}^{*x} f, \quad \text{for every number x in } [r,s],$$

then

$$\overset{*}{D}g = f, \quad \text{on } [r,s].$$

Second Fundamental Theorem of *-Calculus

If $\overset{*}{D}h$ is continuous on a positive interval [r,s], then

$$\int_{r}^{*s} (\overset{*}{D}h) = h(s) / h(r).$$

Just as the Second Fundamental Theorem of Classical Calculus is useful for evaluating classical integrals, the Second Fundamental Theorem of *-Calculus is useful for evaluating *-integrals. For example, let f(x) = exp x on R_{+}. Since $\overset{*}{D}f$

$= f,$

$$\int_r^{*s} f = \int_r^{*s} (\overset{*}{D} f) = f(s) / f(r).$$

2.11 RELATIONSHIP TO THE CLASSICAL CALCULUS

In view of the preceding development in this chapter, it should be clear that the *-calculus is a self-contained system independent of any other system of calculus. Therefore the reader may be surprised to learn that there is a uniform relationship between the corresponding operators of the *-calculus and the classical calculus.

To exhibit that uniform relationship, we shall use the following notations: for each bipositive function f, we set $\bar{f}(t) = \ln(f(\exp t))$, and for each positive number x, we set $\bar{x} = \ln x$. Then we have the following results.

(1) $\quad \overset{*}{G}{}_r^s f = \exp \left\{ G_{\bar{r}}^{\bar{s}} \bar{f} \right\}$,

(2) $\quad [\overset{*}{D} f](a) = \exp \left\{ [D\bar{f}](\bar{a}) \right\}$,

(3) $\quad \overset{*}{A}{}_r^s f = \exp \left\{ A_{\bar{r}}^{\bar{s}} \bar{f} \right\}$,

(4) $\quad \int_r^{*s} f = \exp \left\{ \int_{\bar{r}}^{\bar{s}} \bar{f} \right\}$.

In (2), the indicated derivatives coexist; in (3) and (4), we assume that f is continuous on the positive interval [r,s].

Of course, it is also possible to express each classical operator in terms of the corresponding *-operator.

The foregoing relationships suggest that for each theorem

in classical calculus there is a corresponding theorem in *-
calculus, and conversely. For instance, here is a Mean Value
Theorem of *-Calculus:

> If a bipositive function f is continuous on a posi-
> tive interval [r,s] and *-differentiable everywhere
> between r and s, then between r and s there is a
> number at which the *-derivative of f equals the *-
> gradient of f on [r,s].

CHAPTER 3

Geometric Arithmetic

3.1 INTRODUCTION

> "The laws of number...are not the laws of nature..., they are laws of the laws of nature."
>
> Frege

Classical calculus and Cartesian analytic geometry are based on classical arithmetic, which is usually called the real number system. But it was the use of nonclassical arithmetics that led to the general theory of the non-Newtonian calculi, to the development of non-Cartesian analytic geometries, to the creation of a new theory of subjective probability, and to the conception of new kinds of vectors, centroids, least-squares methods, and complex numbers. Furthermore, nonclassical arithmetics may also be useful in devising new systems of measurement that will yield simpler physical laws. This was clearly recognized by Norman Robert Campbell, a pioneer in the theory of measurement:

> "...we must recognize the possibility that a system of measurement may be arbitrary otherwise than in the choice of unit; there may be arbitrariness in the choice of the process of addition."[1]

In this chapter we shall describe one nonclassical arithmetic, geometric arithmetic, which will be used throughout the remainder of the book.[2]

N O T E S

1.   The quotation is from page 292 of Campbell's remarkable
book entitled *Foundations of Science* (Dover reprint, 1957).

2.   The nonclassical arithmetics and non-Newtonian calculi
should be distinguished from the nonstandard arithmetic and
analysis developed by Abraham Robinson.   (Nonstandard analysis
is really classical calculus developed with a rigorous use of
infinitesimals.)

3.2   CLASSICAL ARITHMETIC

Classical arithmetic has been used for centuries but was

not established on a sound axiomatic basis until the latter

part of the nineteenth century.   However, the details of such

a treatment are not essential here.[1]

Informally, <u>classical arithmetic</u> (or the real number sys-

tem) is a system consisting of a set R, for which there are

four operations +, -, ×, / and an ordering relation <, all

subject to certain familiar axioms commonly referred to as the

complete-ordered-field axioms.   The members of R are called

numbers, and the set R is called the <u>realm</u> of classical arith-

metic.[2]

N O T E S

1.   A new axiomatic treatment of classical arithmetic is presented in detail in [1].   The first full axiomatization of
classical arithmetic was probably given by Hilbert in his article "Uber den Zahlbegriff," published in 1900.   The term
"classical arithmetic" was used by William and Martha Kneale
in *The Development of Logic* (Oxford University Press, 1962).

2.   The reader is reminded that in this book the word "number"
means real number.   The term "realm" was introduced in [2].

3.3   GEOMETRIC ARITHMETIC

By <u>geometric arithmetic</u> we mean the system consisting of the set R_+ of all positive numbers, the usual ordering relation $<$, and four operations $\dot{+}$, $\dot{-}$, $\dot{\times}$, $\dot{/}$ defined for R_+ as follows.

<u>Geometric addition</u>: $a \dot{+} b = \exp(\ln a + \ln b)$

<u>Geometric subtraction</u>: $a \dot{-} b = \exp(\ln a - \ln b)$

<u>Geometric multiplication</u>: $a \dot{\times} b = \exp(\ln a \times \ln b)$

<u>Geometric division</u>: $a \dot{/} b = \exp(\ln a / \ln b)$ if $b \neq 1$

Observe that

$$a \dot{+} b = ab,$$

$$a \dot{-} b = a/b,$$

$$a \dot{\times} b = a^{\ln b} = b^{\ln a},$$

$$a \dot{/} b = a^{1/\ln b} \text{ if } b \neq 1.$$

The <u>realm</u> of geometric arithmetic is R_+.

The fact that geometric arithmetic satisfies all the complete-ordered-field axioms has an important practical implication: the rules for handling geometric arithmetic are exactly the same as the rules for handling classical arithmetic.   For example:

$$a \dot{+} b = b \dot{+} a,$$

$$a \dot{\times} b = b \dot{\times} a,$$

$$a \dot{\times} (b \dot{+} c) = (a \dot{\times} b) \dot{+} (a \dot{\times} c).$$

For each number r, we set $\dot{r} = e^r$.   For example, $\dot{0} = e^0 = 1$ and $\dot{1} = e^1 = e$.

Since

$$y \mathbin{\dot{+}} \dot{0} = y \text{ and } y \mathbin{\dot{\times}} \dot{1} = y$$

for each number y in R_+, we see that $\dot{0}$ and $\dot{1}$ are the "zero" and "one" in geometric arithmetic.[1]

The <u>geometric integers</u> are the numbers $\dot{n}$, where n is an arbitrary integer; if $\dot{0} < \dot{n}$, then

$$\dot{n} = \underbrace{\dot{1} \mathbin{\dot{+}} \cdots \mathbin{\dot{+}} \dot{1}.}_{n \text{ terms}}$$

The <u>geometrically-positive numbers</u> are the numbers in R_+ greater than $\dot{0}$, and the <u>geometrically-negative numbers</u> are the numbers in R_+ less than $\dot{0}$.

For each number p in R_+, we make the following defini-tions:

$$\dot{-}p = \dot{0} \mathbin{\dot{-}} p;$$

$$p^{\dot{2}} = p \mathbin{\dot{\times}} p; \quad \text{(See Note 2.)}$$

$$|\dot{p}| = \begin{cases} p & \text{if } p \geq \dot{0} \\ \dot{-}p & \text{if } p < \dot{0}; \end{cases}$$

if $p \geq \dot{0}$, then $\sqrt{p}$ is the unique number s in R_+ such that $s \geq \dot{0}$ and $s^{\dot{2}} = p$.

It turns out that

$$\dot{-}(\dot{-}p) = p;$$

$$(\sqrt{p})^{\dot{2}} = p, \quad \text{if } p \geq \dot{0};$$

$$\sqrt{p^{\dot{2}}} = |\dot{p}|.$$

Also,

$$\dot{-}p = \exp\{-\ln p\} = 1/p;$$

$$p^{\dot{2}} = \exp\left\{ (\ln p)^2 \right\} = p^{\ln p};$$

$$\dot{|}p\dot{|} = \exp\left\{ |\ln p| \right\};$$

$$\dot{\sqrt{}}\,\overline{p} = \exp\left\{ \sqrt{\ln p} \right\}, \quad \text{if } p \geq \dot{0}.$$

The following comparisons show that the role of the geometric average in geometric arithmetic is analogous to the role of the arithmetic average in classical arithmetic.

Let v be the arithmetic average of n numbers $v_1, \ldots, v_n$, and let w be the geometric average of n positive numbers $w_1, \ldots, w_n$. Then v, which equals

$$(v_1 + \cdots + v_n) / n,$$

is the unique number such that

$$\underbrace{v + \cdots + v}_{n \text{ terms}} = v_1 + \cdots + v_n;$$

and w, which equals

$$(w_1 \dot{+} \cdots \dot{+} w_n) \dot{/} \dot{n},$$

is the unique positive number such that

$$\underbrace{w \dot{+} \cdots \dot{+} w}_{n \text{ terms}} = w_1 \dot{+} \cdots \dot{+} w_n.$$

(Thus, it is appropriate to say that the arithmetic and geometric averages are the "natural" averages of classical and geometric arithmetic, respectively.)

Furthermore,

(1) $(v - v_1) + \cdots + (v - v_n) = 0;$

(2) the expression

$$\sqrt{(x - v_1)^2 + \cdots + (x - v_n)^2},$$

where x is unrestricted in R,

is a minimum when and only when $x = v$;

(3) $(w \mathbin{\dot-} w_1) \mathbin{\dot+} \cdots \mathbin{\dot+} (w \mathbin{\dot-} w_n) = \dot{0}$;

(4) the expression

$$\sqrt{(x \mathbin{\dot-} w_1)^{\dot{2}} \mathbin{\dot+} \cdots \mathbin{\dot+} (x \mathbin{\dot-} w_n)^{\dot{2}}} \, ,$$

where x is unrestricted in R_+,

is a minimum when and only when $x = w$.

It is convenient to conceive the radical expression in (2) above as representing the "classical distance" from x to $v_1, \ldots, v_n$. (For $n = 1$, the expression reduces to $|x - v_1|$, which is the usual distance from x to v_1.) Accordingly, one may say that the arithmetic average of $v_1, \ldots, v_n$ is the number that is "classically closest" to $v_1, \ldots, v_n$.

Similarly, it is convenient to conceive the radical expression in (4) above as representing the "geometric distance" from x to $w_1, \ldots, w_n$. (For $n = 1$, the "geometric distance" equals $|x \mathbin{\dot-} w_1|$.) Thus one may say that the geometric average of $w_1, \ldots, w_n$ is the positive number that is "geometrically closest" to $w_1, \ldots, w_n$.

The following two facts show that the geometric progressions play the same role in geometric arithmetic as the arithmetic progressions do in classical arithmetic.

An arithmetic progression is a finite sequence of numbers $v_1, \ldots, v_n$ such that $v_i - v_{i-1}$ is the same for every integer i from 2 to n.

A geometric progression is a finite sequence of positive numbers $w_1, \ldots, w_n$ such that $w_i \doteq w_{i-1}$ is the same for every integer i from 2 to n.

It should be quite clear that every concept in classical arithmetic has a counterpart in geometric arithmetic, and conversely.

Obviously geometric arithmetic is especially useful in situations where products and ratios provide the natural methods of combining and comparing magnitudes. Although geometric arithmetic applies only to positive numbers, it is possible to construct a geometric-type arithmetic that applies to negative numbers.[3]

<u>N O T E S</u>

1. The stipulation in Section 2.9 that $\int_{r}^{*r} f = 1 = \dot{0}$ clearly parallels the stipulation in Section 1.9 that $\int_{r}^{r} f = 0$.

Furthermore, our reason for using e rather than some other positive constant in our definition of *-slope (Section 2.3) stems from the fact that e is the "one" of geometric arithmetic. That definition may be restated as follows.
        The <u>*-slope</u> of a power function p is the positive number $\overline{p(b)} \doteq p(a)$, where a and b are any two positive numbers such that $b \doteq a = \dot{1}$.
The reader may wish to compare that with the definition of classical slope in Section 1.3.

2. Since $\dot{2} = e^2$, there is a slight risk that the reader will take $p^{\dot{2}}$ to be $p^{(e^2)}$. We wish to stress that $p^{\dot{2}}$ is defined to be $p \overset{\cdot}{\times} p$, which equals $p^{\ln p}$.

3. See Section 5.4 of [2].

3.4 COMPARISON OF THE CLASSICAL AND BIGEOMETRIC CALCULI

First we wish to point out that the power functions, which are the standards of comparison in the *-calculus, may be described in the following manner, thus exhibiting their similarity to the linear functions, which are the standards of comparison in the classical calculus.

A power function is any bipositive function p on R_+ such that $p(x) = (m \overset{.}{\times} x) \overset{.}{+} c$, where m and c are positive constants.[1]

By using geometric arithmetic to express the operators of the *-calculus, we shall reveal some similarities between the classical operators and the *-operators.

Gradients

$$G_r^s f = [f(s) - f(r)] / (s - r),$$

$$\overset{*s}{G}_r f = [f(s) \overset{.}{-} f(r)] \overset{.}{/} (s \overset{.}{-} r).$$

Derivatives

$$[Df](a) = \lim_{x \to a} \left\{ [f(x) - f(a)] / (x - a) \right\},$$

$$[\overset{*}{D}f](a) = \lim_{x \to a} \left\{ [f(x) \overset{.}{-} f(a)] \overset{.}{/} (x \overset{.}{-} a) \right\}.$$

Averages

If $a_1, \ldots, a_n$ is the n-fold arithmetic partition of $[r,s]$, then

$$A_r^s f = \lim_{n \to \infty} \left\{ [f(a_1) + \cdots + f(a_n)] / n \right\}.$$

If $a_1, \ldots, a_n$ is the n-fold geometric partition of $[r,s]$, then

$$\overset{*s}{A}_r f = \lim_{n \to \infty} \left\{ [f(a_1) \overset{.}{+} \cdots \overset{.}{+} f(a_n)] \overset{..}{/} n \right\}.$$

Integrals

$$\int_r^s f = (s - r) \times A_r^s f,$$

$$\overset{*}{\int_r^s} f = (s \overset{\cdot}{-} r) \overset{\cdot}{\times} \overset{*}{A}_r^s f.$$

It was noted in Section 2.4 that the *-gradient is multiplicative, divisional, and involutional:

$$\overset{*}{G}_r^s (f \cdot g) = \overset{*}{G}_r^s f \cdot \overset{*}{G}_r^s g,$$

$$\overset{*}{G}_r^s (f / g) = \overset{*}{G}_r^s f / \overset{*}{G}_r^s g,$$

$$\overset{*}{G}_r^s (f^c) = (\overset{*}{G}_r^s f)^c, \quad c \text{ any constant.}$$

By using geometric arithmetic to re-express those three properties, we find that they are actually conditions of additivity, subtractivity, and homogeneity within geometric arithmetic:

$$\overset{*}{G}_r^s (f \overset{\cdot}{+} g) = \overset{*}{G}_r^s f \overset{\cdot}{+} \overset{*}{G}_r^s g,$$

$$\overset{*}{G}_r^s (f \overset{\cdot}{-} g) = \overset{*}{G}_r^s f \overset{\cdot}{-} \overset{*}{G}_r^s g,$$

$$\overset{*}{G}_r^s (c \overset{\cdot}{\times} f) = c \overset{\cdot}{\times} \overset{*}{G}_r^s f, \quad c \text{ any constant.}$$

The corresponding properties of the operators $\overset{*}{D}$, $\overset{*}{A}$, and $\overset{*}{\int}$ may be re-expressed similarly.

N O T E

1. Another way of using geometric arithmetic to exhibit the similarity between the power functions and the linear functions is as follows.
Define the classical length of an interval [r,s] to be s - r, and define the classical change of a function f on [r,s] to be f(s) - f(r). Then, according to Section 1.2, each linear function is uniform in the sense that it has the same classical change on any two intervals with the same classical length.

Now, define the <u>geometric length</u> of a positive interval
[r,s] to be s $\doteq$ r, and define the <u>geometric change</u> of a bipositive function f on [r,s] to be f(s) $\doteq$ f(r). Then, according
to Section 2.2, each power function is uniform in the sense
that it has the same geometric change on any two positive intervals with the same geometric length.

3.5 ARITHMETICS AND CALCULI

An <u>arithmetic</u> is any system that satisfies the complete-
ordered-field axioms and has a realm that is a subset of R.
There are infinitely-many arithmetics, all of which are isomorphic, that is, structurally equivalent. Nevertheless the
fact that two systems are isomorphic does *not* preclude their
separate uses.

In [2], it is shown that each ordered pair of arithmetics
gives rise to a calculus by a judicious use of the first a-
rithmetic for function arguments and the second arithmetic for
function values. The following chart indicates the four calculi obtainable by using the classical and geometric arithmetics.

	1st arithmetic (arguments)	2nd arithmetic (values)
classical calculus:	classical	classical
bigeometric calculus:	geometric	geometric
geometric calculus:	classical	geometric
anageometric calculus:	geometric	classical

The geometric calculus is treated in detail in [3], where it is called the exponential calculus. In [2], the geometric, anageometric, and bigeometric calculi are treated briefly.[1]

Also of interest are the harmonic and quadratic arithmetics, which, when used in tandem with classical arithmetic, give rise to the following calculi:

	1st arithmetic (arguments)	2nd arithmetic (values)
harmonic calculus:	classical	harmonic
biharmonic calculus:	harmonic	harmonic
anaharmonic calculus:	harmonic	classical
quadratic calculus:	classical	quadratic
biquadratic calculus:	quadratic	quadratic
anaquadratic calculus:	quadratic	classical

In [2], the foregoing calculi are discussed briefly.[2]

Of course, one may use any two arithmetics, e.g., geometric arithmetic for arguments and quadratic arithmetic for values.

N O T E S

1. The geometric gradient is related to the so-called "compound growth rate" used in securities analysis and elsewhere; the geometric derivative is related to the well-known logarithmic derivative; and the geometric average fits naturally into the scheme of geometric calculus. The anageometric derivative turns out to be the classical derivative with respect to ln; and the anageometric integral is of the Stieltjes variety.

2. The harmonic progressions arise naturally in harmonic arithmetic. The harmonic and quadratic averages play important roles in the harmonic and quadratic calculi, respectively.

CHAPTER 4

Graphical Interpretations

4.1 INTRODUCTION

> "Perhaps the one tendency that did more than any
> other to conceal from mathematicians for almost two
> centuries the logical basis of the [classical] cal-
> culus was the result of the attempt to make geome-
> trical, rather than arithmetic, conceptions funda-
> mental."
>
> Carl B. Boyer[1]

Originally we conceived the non-Newtonian calculi ana-

lytically, not geometrically. Indeed, only at a relatively

late stage of our investigations were we able to interpret the

operators of the non-Newtonian calculi in a geometric manner

that we considered to be suitable. (The suitability of such

interpretations is of course a subjective matter.) The dis-

covery of those geometric interpretations ultimately led to

the development of non-Cartesian geometries, one of which will

be discussed in Section 6.3.

N O T E

1. Carl B. Boyer, *The History of the Calculus and Its Conceptual Development* (New York: Dover reprint, 1949), p. 104.

47

4.2 BIGEOMETRIC GRAPHS

By *-paper we mean paper that is ruled off in squares and labeled as follows:

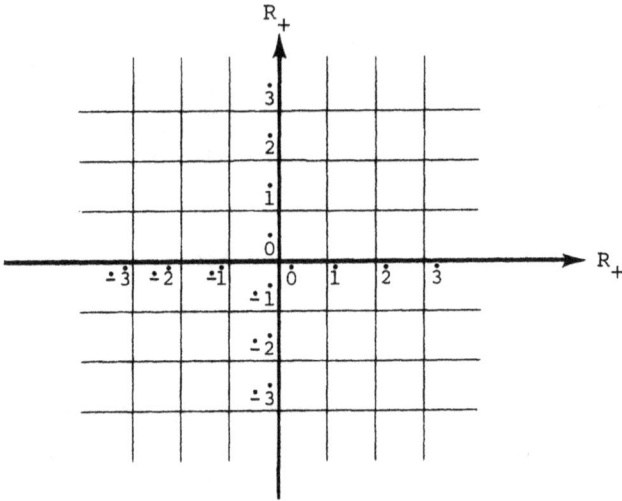

On each axis, the indicated equi-spaced points are marked with the geometric integers

$$\ldots, \ ^{\pm}\dot{2}, \ ^{\pm}\dot{1}, \ \dot{0}, \ \dot{1}, \ \dot{2}, \ \ldots,$$

which equal

$$\ldots, \ e^{-2}, \ e^{-1}, \ e^{0}, \ e^{1}, \ e^{2}, \ \ldots.$$

The origin corresponds to $(\dot{0}, \dot{0})$.

Although *-paper is equivalent to so-called log-log paper, the former is better suited for our purposes.

The *-graph of a set of bipositive points is the result of plotting all its members on *-paper. Of course, only bipositive points can be plotted on *-paper.

4.3 BIGEOMETRIC DISTANCE

For convenience the symbol P_i will henceforth be used to denote the point (x_i, y_i).

The *-distance between bipositive points P_1 and P_2 is denoted by $\overset{*}{d}(P_1, P_2)$ and is defined to be the number

$$\overset{\bullet}{\sqrt{}}(x_1 \overset{\bullet}{-} x_2)^{\overset{\bullet}{2}} \overset{\bullet}{+} (y_1 \overset{\bullet}{-} y_2)^{\overset{\bullet}{2}} \ .$$

This equals

$$\exp\left\{ \sqrt{(\ln x_1 - \ln x_2)^2 + (\ln y_1 - \ln y_2)^2} \right\}$$

and may be obtained graphically by plotting P_1 and P_2 on *-paper and measuring their separation with the "ruler" provided by either axis.[1]

It turns out that

$$\overset{*}{d}(P_1, P_2) \geq \overset{\bullet}{0};$$

if $x_1 = x_2$, then $\overset{*}{d}(P_1, P_2) = \overset{\bullet}{|}y_1 \overset{\bullet}{-} y_2\overset{\bullet}{|}$; and

if $y_1 = y_2$, then $\overset{*}{d}(P_1, P_2) = \overset{\bullet}{|}x_1 \overset{\bullet}{-} x_2\overset{\bullet}{|}$.

Furthermore, $\overset{*}{d}$ is multiplicative along power functions; that is, for any power function p and any bipositive points P_1, P_2, and P_3 in p such that $x_1 \leq x_2 \leq x_3$,

$$\overset{*}{d}(P_1, P_2) \cdot \overset{*}{d}(P_2, P_3) = \overset{*}{d}(P_1, P_3).$$

In Section 6.3, *-distance will be discussed further.

N O T E

1. We have never seen *-distance in the literature.

4.4 GRAPHICAL INTERPRETATION OF BIGEOMETRIC SLOPE

As expected, the *-graph of each power function is a
straight line.

Consider any power function p whose *-slope is greater
than $\overset{\bullet}{0}$. Choose any two distinct points P_1 and P_2 on p, and
let P_3 be as indicated in the following figure.

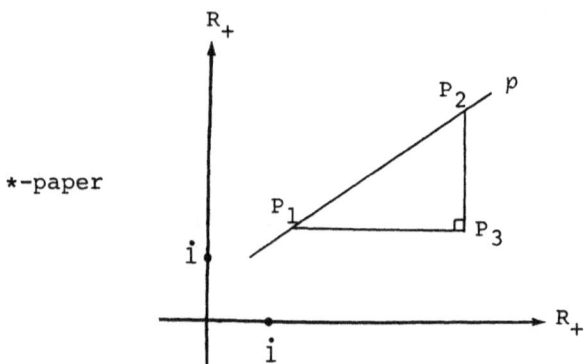

Then the *-slope of p equals $\overset{*}{d}(P_2,P_3) \, / \, \overset{\bullet\,*}{d}(P_1,P_3)$.

With obvious adjustments an interpretation can be given
if the *-slope of p is less than $\overset{\bullet}{0}$, in which case the *-graph
of p is decreasing. If the *-slope of p equals $\overset{\bullet}{0}$, then the
*-graph of p is horizontal.

It is now easy to formulate a graphical interpretation
of the *-gradient, since it is defined directly in terms of
*-slope. We omit the details.

4.5 GRAPHICAL INTERPRETATION OF THE BIGEOMETRIC DERIVATIVE

In Section 2.5, it was observed that, if it exists, the
*-derivative of a bipositive function f at an argument a is
equal to the *-slope of the power function p that is tangent
to f at (a,f(a)). The following figure illustrates the situ-
ation.

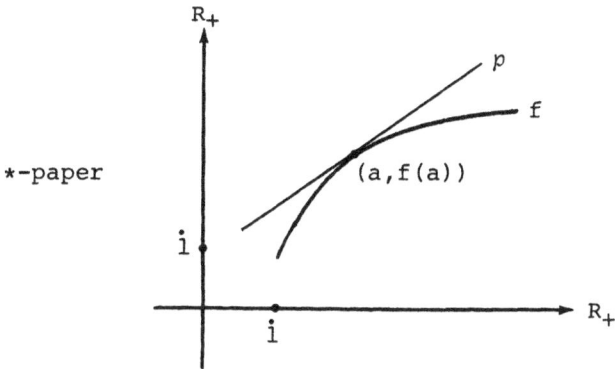

4.6 GRAPHICAL INTERPRETATION OF THE BIGEOMETRIC INTEGRAL

In this section all references to geometric figures are
intended to apply to figures as they appear on *-paper.

A *-unit square is a square with sides of *-length $\dot{1}$.

The *-area of a rectangle is the geometric product of its
*-length and *-width. Of course, the *-area of a *-unit
square is $\dot{1}$. If a rectangle is decomposable into n *-unit
squares, then its *-area equals $\dot{n}$, as expected.

Let f be a continuous bipositive function on a positive

interval [r,s] such that all values of f are greater than $\overset{\cdot}{0}$. Let S be the region bounded by the *-graph of f, the horizontal axis, and the vertical lines at r and s.

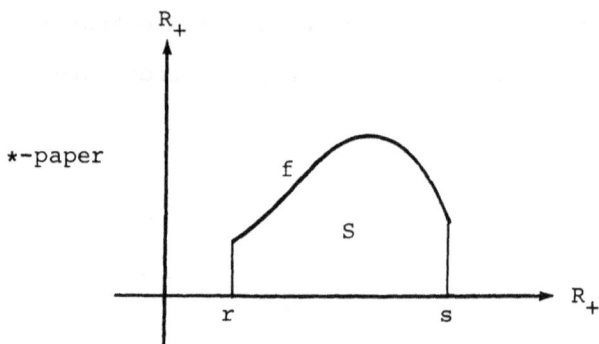

By partitioning the region S into approximating rectangles and using the customary limit process, one can readily define the *-area of S, which turns out to be $\overset{*}{\int_r^s} f$, thereby providing a simple interpretation of the *-integral of f on [r,s].

4.7 GRAPHICAL INTERPRETATION OF THE BIGEOMETRIC AVERAGE

In this section all references to geometric figures are intended to apply to figures as they appear on *-paper.

Let f and S be as indicated in the preceding section, and let m denote $\overset{*}{A}{}_r^s f$. It turns out that m is greater than $\overset{\cdot}{0}$, and that the *-area of S is equal to the *-area of the rectangle indicated in the following figure.

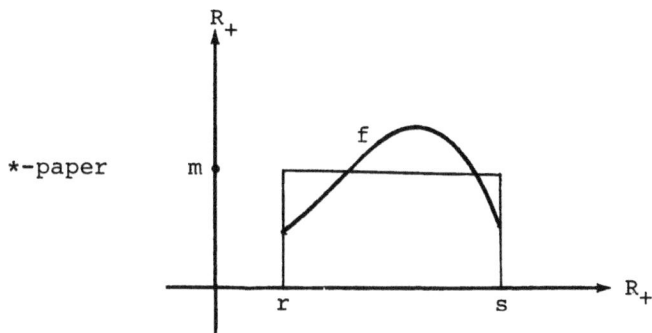

Thus, $\overset{*}{A}{}^{s}_{r}f$ is the *-height of the rectangle in the upper half-plane whose *-base is the line segment from $(r,\overset{\bullet}{0})$ to $(s,\overset{\bullet}{0})$ and whose *-area is equal to the *-area of S.

CHAPTER 5

Heuristic Principles of Application

5.1 INTRODUCTION

> "...all the ingenious analysis which has evolved
> from the hypothesis of linearity is at best a
> first approximation to the applicable mathematics
> of the future."
>
> E. T. Bell[1]

Since there are situations where success in applying the
classical calculus seems to depend on a felicitous choice of
simplifying assumptions, it is reasonable to suppose that
there are situations where the *-calculus might prove to be
useful. Accordingly, this chapter contains some heuristic
principles that may be helpful for selecting appropriate operators from the classical calculus and the *-calculus.

Although one is always free to use any operator that is
meaningful in a given context, a suitable choice of an operator or calculus depends chiefly upon its intended use.

N O T E

1. E. T. Bell, *The Development of Mathematics*, 2nd ed. (New
York: McGraw-Hill, 1945).

5.2 CHOOSING GRADIENTS AND DERIVATIVES

We shall give three heuristic principles for choosing
gradients and derivatives.

<u>Principle I</u>

> Where changes in arguments and values are naturally
> measured by differences, the classical gradient and
> classical derivative should probably be used. Where
> changes in arguments and values are naturally mea-
> sured by ratios, the *-gradient and *-derivative
> should probably be used.[1]

For example, the classical gradient and derivative are
appropriate for analyzing rectilinear motion because changes
in time and position are best measured by differences. But
the *-gradient and *-derivative are appropriate for analyzing
the growth of one organ relative to the growth of another organ (in the same body) because growth-changes are best measured by ratios.

<u>Principle II</u>

> Where the relationship between two magnitudes is, or
> is assumed to be, a linear function under normal or
> ideal conditions, the classical gradient and classi-
> cal derivative should probably be used. Where the
> relationship between the two magnitudes is, or is
> assumed to be, a power function under normal or i-
> deal conditions, the *-gradient and *-derivative
> should probably be used.[2]

For instance, the classical gradient and classical derivative are useful for studying rectilinear motion because under ideal conditions, that is, the absence of forces, the relationship of position to time is assumed to be linear (Newton's First Law). On the other hand, the *-gradient and *-derivative ought to be useful in psychophysics because the relationship of sensation to stimulus normally follows a power law.

Before we state Principle III, it will be helpful to repeat some earlier definitions and results.

If each point (x,y) of a function is changed to (x + a, y + b), where a and b are constants, then a change of origins has been made in the function. It turns out that the classical gradient and classical derivative are origin-free, that is, invariant under every change of origins.

If each point (x,y) of a bipositive function is changed to (px, qy), where p and q are positive constants, then a change of scales (or units) has been made in the function. It turns out that the *-gradient and *-derivative are scale-free, that is, invariant under every change of scales.

Principle III

> If one desires an origin-free gradient or an origin-free derivative, then he should probably choose the classical gradient or classical derivative. If one desires a scale-free gradient or a scale-free derivative, then he should probably choose the *-gradient or *-derivative.

For example, physicists usually desire that the average velocity of an object moving rectilinearly be independent of the origins used for measuring time and position. Accordingly, they define the average velocity of the object to be the classical gradient of its position function.

However, a physicist who desires an average velocity concept that is independent of the scales (or units) used for measuring time and position could define average velocity to be the *-gradient of the position function.

It is quite possible that scale-free gradients and scale-free derivatives will be of considerable interest to physicists, in view of the following remarks by John A. Wheeler:

> "In the past five years one of the greatest developments in elementary particle physics has been so-called conformal invariance, the discovery that the equations of elementary particle physics possess a property that in effect is this: changing the scale in which you examine the phenomenon does not change the nature of the phenomenon. For decades this so-called conformal invariance has been known to be a property of electromagnetism. The conformal feature of gravitational physics has been known for a lesser time. But the conformal features of elementary particle physics are a discovery of the past five years. This strange feature of nature that permits us to subsume the very large and the very small under the same kind of equation is something we don't yet understand, but it raises in one's mind the perpetual question: Can it be true that what we think of as the very small and what we think of as the very large are really not so different?"[3]

N O T E S

1. Where changes in arguments and values are naturally measured by differences and ratios respectively, the geometric gradient and geometric derivative, which are discussed in [2] and [3], should probably be used. Where changes in arguments

and values are naturally measured by ratios and differences respectively, the anageometric gradient and anageometric derivative, which are discussed in [2], should probably be used.

2. Where the relationship is, or is assumed to be, an exponential function under normal or ideal conditions, the geometric gradient and geometric derivative should probably be used.

3. *Intellectual Digest*, June 1973.

5.3 CHOOSING INTEGRALS

Suppose that a scientific relationship has been expressed by a differential equation involving *-derivatives and perhaps operations of geometric arithmetic. To solve that equation one would certainly try to use *-integration. If obtained, the solutions would initially be expressed by means of *-integrals with variable upper limits, which in turn would be evaluated in terms of known functions or new functions defined for the purpose. (Many functions were originally defined as classical integrals with variable upper limits.) It is even conceivable that solutions in closed form could be obtained for certain intractable differential equations by re-expressing them with *-derivatives and operations of geometric arithmetic.

A differential equation containing classical derivatives and *-derivatives can readily be transformed into an equation involving only classical derivatives or only *-derivatives. (See Section 2.11.)

Integrals are also used for *defining* scientific concepts.

Consider, for example, the concept of work. Suppose a point mass moves along the x-axis as a result of a force directed along the axis. Let the force be f(x) at position x. If the force f is constant, the work on the position interval [r,s] is defined to be (s - r) · f, which we denote by $W_r^s f$. When f is constant the following conditions are clearly satisfied.

1. $W_r^s f = (s - r) \cdot f$.

2. Work is monotonically increasing with respect to force.

3. Work is additive with respect to position; that is, for any positions r, s, t such that r < s < t,

    $$W_r^s f + W_s^t f = W_r^t f.$$

Desiring to extend the concept of work to the case where the force f is continuously variable, the physicist stipulates that the extended work concept should satisfy Condition 1 when f is constant and should satisfy Conditions 2 and 3 in gener- al. The solution is now uniquely determined; for according to Section 1.9, the operator W satisfies those three conditions if and only if W is the classical integral. Thus, the physicist must adopt the following definition:

$$W_r^s f = \int_r^s f.$$

The foregoing well-known example illustrates the fact that the characterization of the classical integral in Section 1.9 is a heuristic principle for its appropriate use.

Now suppose we are concerned with a positive magnitude g, called "gorce," that may be continuously variable with respect to a positive magnitude called "bosition." If g is constant, we define the "toil" on the bosition interval [r,s] to be $g^{\ln s - \ln r}$, denoted by $T_r^s g$.[1] When g is constant, the following conditions are clearly satisfied.

1. $T_r^s g = g^{\ln s - \ln r}$.

2. Toil is monotonically increasing with respect to gorce.

3. Toil is multiplicative with respect to bosition; that is, for any bositions r, s, t such that r < s < t,

$$T_r^s g \cdot T_s^t g = T_r^t g.$$

To extend the toil concept to the case where the gorce g is continuously variable, we stipulate that the extended toil concept should satisfy Condition 1 when g is constant and should satisfy Conditions 2 and 3 in general. The solution is now uniquely determined; for, according to Section 2.9, the operator T satisfies those three conditions if and only if T is the ∗-integral. Thus, the following definition must be adopted:

$$T_r^s g = \int_r^{*s} g.$$

It should now be clear that the characterization of the ∗-integral in Section 2.9 is a heuristic principle for its appropriate use.

N O T E

1. Notice that $g^{\ln s - \ln r}$ equals $(s \overset{\cdot}{-} r) \overset{\cdot}{\times} g$, an expression that conceivably could arise naturally in some scientific application of geometric arithmetic.

5.4 CHOOSING AVERAGES

Most of this section is devoted to averages of finite sequences of numbers rather than to averages of continuous functions, since a choice of the latter depends on a choice of the former.[1]

We are not concerned here with probabilistic justifications for the use of a particular average. Nevertheless, we believe that any such justification for the use of the arithmetic average relative to classical arithmetic can be matched by a similar justification for the use of the geometric average relative to geometric arithmetic.

Historically one reason for the popularity of the arithmetic average is its simplicity of calculation, but that issue is surely irrelevant in this age of computers. For instance, one investment advisory service had for many years maintained a geometric average of 1000 stocks.[2]

In choosing a method of averaging physical magnitudes the fundamental issue is the natural method of combining them. Where physical magnitudes are naturally combined by addition (multiplication), the arithmetic average (geometric average) is physically meaningful and possibly useful. One should keep in mind that if magnitudes m_1 and m_2 are naturally combined by addition, then the magnitudes p^{m_1} and p^{m_2} are naturally combined by multiplication (p a positive constant).

We shall let Au_i represent the arithmetic average of n numbers $u_1,\ldots,u_n$, and if the u_i are positive we shall let $\dot{A}u_i$ represent their geometric average.

Since the geometric and arithmetic averages are the "natural" averages in geometric and classical arithmetics respectively (Section 3.3), the geometric average has the same properties relative to geometric arithmetic as the arithmetic average has relative to classical arithmetic. For example, since

$$(1) \quad A(u_i - v_i) = Au_i - Av_i,$$

one should expect that

$$\dot{A}(u_i \div v_i) = \dot{A}u_i \div \dot{A}v_i.$$

And indeed, the preceding equation is valid since it is merely a statement of the divisional character of $\dot{A}$:

$$(2) \quad \dot{A}(u_i / v_i) = (\dot{A}u_i) / (\dot{A}v_i).$$

Items (1) and (2) above are often cited to rationalize the heuristic principle that differences are best averaged arithmetically, but that ratios are best averaged geometrically. However, there are situations where the arithmetic average of ratios is significant.[3]

Consider the problem of estimating the area of a rectangle from n measurements $x_1,\ldots,x_n$ of its length and n measurements $y_1,\ldots,y_n$ of its width. The following four estimates will be considered.

I. $E_1 = (Ax_i) \cdot (Ay_i)$

II. $E_2 = A(x_i \cdot y_i)$

III. $E_3 = (\dot{A}x_i) \cdot (\dot{A}y_i)$

IV. $E_4 = \dot{A}(x_i \cdot y_i)$

Because $E_1 \neq E_2$ except in isolated cases, and because $E_3 = E_4$, it would appear that here the geometric average is more appropriate than the arithmetic average. But that is to be expected since the geometric average is multiplicative and the area is the result of multiplication. (However, a similar analysis would indicate that the arithmetic average would be more appropriate than the geometric average if one were estimating the perimeter of the rectangle.) Furthermore, Method I, which is quite popular with scientists, has another disconcerting feature: If there arose new measurements x_{n+1} and y_{n+1} such that $x_{n+1} \cdot y_{n+1} = E_1$, then Method I applied to $x_1, \ldots, x_n, x_{n+1}$ and $y_1, \ldots, y_n, y_{n+1}$ *does not* yield the original estimate E_1 except in trivial cases. On the other hand, if there arose new measurements x_{n+1} and y_{n+1} such that $x_{n+1} \cdot y_{n+1} = E_3 = E_4$, then Methods III and IV applied to $x_1, \ldots, x_n, x_{n+1}$ and $y_1, \ldots, y_n, y_{n+1}$ *do* yield the original estimate E_3 ($= E_4$).

Now consider the problem of estimating the density of an object, given n measurements of its mass $m_1, \ldots, m_n$ and n measurements of its volume $v_1, \ldots, v_n$. There are at least four estimates of the density worthy of consideration.

$E_1 = (Am_i) / (Av_i)$

$$E_2 = A(m_i \,/\, v_i)$$
$$E_3 = (\dot{A}m_i) \,/\, (\dot{A}v_i)$$
$$E_4 = \dot{A}(m_i \,/\, v_i)$$

Since $E_1 \neq E_2$ except in isolated cases, and since $E_3 = E_4$, it seems that the choice ought to be the geometric average.

Scientists often invoke a "normalcy" principle for making an appropriate choice of average for continuous functions on intervals. For example, it is widely accepted that the arithmetic average is appropriate where the functional relationship would be linear under normal conditions. Presumably that idea arose from the observation that the arithmetic average of a linear function on [r,s] is equal to its value at the arithmetic average of r and s, and is also equal to the arithmetic average of its values at r and s. By the same token the *-average is appropriate where under normal conditions the functional relationship would be of the power-function variety, since the *-average of a power function on a positive interval [r,s] is equal to its value at the geometric average of r and s, and is also equal to the geometric average of its values at r and s.

Some scientists, notably the psychophysicist S. S. Stevens of Harvard University, favor the use of certain invariance principles for choosing averages.[4]

As in the case for integrals, heuristic principles for the use of averages are also provided by their characterizations (Sections 1.6 and 2.6).

In many situations, averages are intuitively more satisfying than integrals. Consider, for instance, a particle moving rectilinearly with positive velocity v. The distance s
traveled in the time interval [a,b] is given by

$$s = \int_a^b v.$$

Although that fact may be clear to a student, he may nevertheless find that the following formula conveys a more immediate
meaning:

$$s = (b - a) \cdot A_a^b v;$$

that is, the distance traveled equals the product of the time
elapsed and the arithmetic average of the velocity. This version is a direct extension of the case where v is constant.
Other examples will readily occur to the reader. (For the
"toil" concept in the preceding section, we have

$$T_r^s g = (A_r^s g)^{\ln s - \ln r},$$

a formula that is a direct extension of the case where g is
constant.)

N O T E S

1. It also depends on a choice of the method of partitioning
argument intervals.

2. American Investors Service (Greenwich, Connecticut) dis-
tributed an interesting booklet by George A. Chestnut, Jr.,
who gave some excellent reasons why he considered geometric
averaging the best method of averaging stock prices.

3. For example, suppose that initially $1000 is invested in
one stock at $10 per share and $1000 in another stock at $20
per share. Subsequently the stocks are worth $5 and $50 per

share respectively. Since the original investment of $2000
increased in value to $3000, the overall ratio change in value
is 1.5, which equals the *arithmetic* average of the ratio
changes, 0.5 and 2.5, for the individual stocks.

4. A detailed discussion is given by S. S. Stevens in his article "On the Averaging of Data," which appeared in *Science*,
Volume 121 (January 28, 1955), pp. 113-6. Some comments on
Stevens' ideas may be found in Brian Ellis' book, *Basic Concepts of Measurement* (Cambridge University Press, 1966).

5.5 CONSTANTS AND SCIENTIFIC CONCEPTS

> "Where there is change, he [the scientist] looks
> for constancy in the rate of change; and failing
> that, for constancy in the rate of the rate of
> change."
>
> Nelson Goodman[1]

In this section we illustrate the thesis that the inven-
tion of a scientific concept may depend on the isolation or
discovery of a suitable constant; and we suggest that new scientific concepts may arise from the constants provided by the
*-slopes of power functions.

Consider the concept of average speed. The definition
"distance traveled per unit time" is incomplete because it
fails to provide a method of determining the average speed of
an accelerated particle. The definition "distance divided by
time," though not incorrect, is a gross oversimplification
that fails to reveal the underlying issues. Fortunately there
is a completely satisfactory definition, which undoubtedly was
known to Galileo.

Let us begin with Galileo's definition of <u>uniform motion</u>

as "one in which the distances traversed...during *any* equal intervals of time are themselves equal." (Continuity of the motion is tacitly intended here.)[2] Then we isolate a constant in each given uniform motion by defining <u>speed</u> to be the distance traveled in any unit time-interval. Finally, for a particle that moves non-uniformly a distance d in time t, we define the <u>average speed</u> to be the speed that a particle in uniform motion must have in order to travel a distance d in time t. In our opinion, neither the simplicity nor the obviousness of the answer, d/t, justifies its use as the definition of average speed.

Although Archimedes undoubtedly knew, in effect, the definition of speed in uniform motion, it was probably Galileo who first introduced the concepts of speed and acceleration in non-uniform motion, and he may have been the first to define uniform motion rigorously.

The critical step in defining average speed is the isolation of a constant (speed) in the phenomenon of uniform motion. Similarly, the critical step in defining the *-gradient is the identification of a constant (*-slope) for any given power function.

Wherever a scientific phenomenon, or an idealized version thereof, is describable by a power function, one automatically has a fundamental constant, the *-slope, that may prove to be useful.

For example, consider a fixed light source radiating

light uniformly in all directions. Since the intensity of illumination is inversely proportional to the square of the distance from the source, the relationship of intensity to distance is describable by a power function, whose *-slope is independent of the units used for measuring distance and intensity.

N O T E S

1. Nelson Goodman, *Problems and Projects* (Bobbs-Merrill, 1972), p. 351.

2. This is the first formal definition in Galileo's masterpiece *Two New Sciences*, "Third Day," translated by Henry Crew and Alfonso de Salvio. We have italicized the word "any" for emphasis, particularly since Galileo adds a Caution on this very point. Galileo uses a similar technique to define uniformly accelerated motion.

CHAPTER 6

A Non-Cartesian Geometry

6.1 INTRODUCTION

"In itself, no curve is simpler than another."
Charles S. Peirce[1]

It is well-known that for Euclidean plane geometry one can construct an arithmetic model in which every linear function represents a Euclidean line. In this chapter we shall show how one can construct an arithmetic model for Euclidean plane geometry in which every power function represents a Euclidean line.

The construction of arithmetic models for geometry is of more than purely mathematical interest, as the following remark by the physicist John L. Synge suggests:

> "The relationship of geometry to relativity is most satisfactorily established when geometry is regarded analytically, a 'point' being nothing but a [sequence] of numbers...and a 'line' a set of points. This is a most fruitful way to look at geometry..."[2]

Since we are concerned here solely with Euclidean geometry of the plane, adjectives such as "plane" and "two-dimensional" will be omitted. Nevertheless, our developments can be extended to higher dimensions.

The symbol P_i will still be used to represent the point (x_i, y_i).

N O T E S

1. *The Monist* (January 1891).

2. John L. Synge, *Relativity: The Special Theory*, 2nd ed. (North-Holland Publishing Co., 1964), p. 4.

6.2 CARTESIAN GEOMETRY

In his classic *Grundlagen der Geometrie* (1899), Hilbert constructed for Euclidean geometry an arithmetic model that he called "the ordinary analytic Cartesian geometry," here abbreviated to "Cartesian geometry." Though Hilbert's construction is elegant, a different approach will be used in this section.[1]

Cartesian geometry is the system consisting of the set of all points (i.e. ordered pairs of numbers) in which the classical distance $d(P_1, P_2)$ between points P_1 and P_2 is stipulated to be the nonnegative number

$$\sqrt{(x_1 - x_2)^2 + (y_1 - y_2)^2} \ .$$

Within Cartesian geometry one can define counterparts of all Euclidean notions, some of which will be discussed in this section.

A classical translation is a mapping of each point (x,y) to the point $(x + a, y + b)$, where a and b are constants.[2] It turns out that classical distance is classically invariant; that is, the classical distance between points is unchanged under all classical translations.

The classical-distance operator d is a metric because of

the following result.

Theorem

For all points P_1, P_2, and P_3,

$$d(P_1,P_2) \geq 0,$$

$$d(P_1,P_2) = 0 \text{ if and only if } P_1 = P_2,$$

$$d(P_1,P_2) = d(P_2,P_1),$$

$$d(P_1,P_2) + d(P_2,P_3) \geq d(P_1,P_3).$$

The points P_1, P_2, and P_3 are <u>collinear</u> provided that at least one of the following holds:

$$d(P_2,P_1) + d(P_1,P_3) = d(P_2,P_3),$$

$$d(P_1,P_2) + d(P_2,P_3) = d(P_1,P_3),$$

$$d(P_1,P_3) + d(P_3,P_2) = d(P_1,P_2).$$

A <u>line</u> is a set L of at least two distinct points such that for all distinct points P_1 and P_2 in L, a point P_3 is in L if and only if P_1, P_2, and P_3 are collinear.

A set of points is <u>vertical</u> provided that all its members have the same first coordinate.

Theorem

The class of nonvertical lines is identical with the class of linear functions.

In view of the preceding theorem one may say that Cartesian geometry is a linear model of Euclidean geometry.

Two lines are <u>parallel</u> provided they are identical or have no common point.

Theorem

Two nonvertical lines are parallel if and only if

they have the same classical slope.

A line L_1 is <u>perpendicular</u> to a line L_2 provided they have a unique common point P, and for any points P_1 on L_1 and P_2 on L_2, distinct from P,

$$d(P_1,P) < d(P_1,P_2).$$

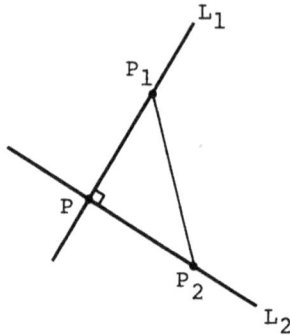

Theorem

Two nonvertical lines are perpendicular if and only if the product of their classical slopes is -1.

Theorem

If two perpendicular lines L_1 and L_2 intersect at point P, then for all points P_1 on L_1 and P_2 on L_2,

$$[d(P_1,P)]^2 + [d(P,P_2)]^2 = [d(P_1,P_2)]^2.$$

N O T E S

1. Hilbert's construction is neatly summarized in Howard Eves' book, *A Survey of Geometry*, Vol.II (Allyn and Bacon, 1965), pp. 102-5. Hilbert's objective was to prove that Euclidean geometry is consistent with classical arithmetic, but that is not our concern here.

2. Clearly, applying a classical translation to the points in a function is equivalent to making a change of origins in the function.

6.3 A NON-CARTESIAN GEOMETRY

> "...the distance concept is logically arbitrary."
> Einstein

By *-geometry we mean the system consisting of the set of all bipositive points (i.e. ordered pairs of positive numbers) in which the *-distance $\overset{*}{d}(P_1, P_2)$ between bipositive points P_1 and P_2 is stipulated to be the number

$$\overset{\bullet}{\sqrt{(x_1 \overset{\bullet}{-} x_2)^{\overset{\bullet}{2}} \overset{\bullet}{+} (y_1 \overset{\bullet}{-} y_2)^{\overset{\bullet}{2}}}} ,$$

which equals

$$\exp \left\{ \sqrt{(\ln x_1 - \ln x_2)^2 + (\ln y_1 - \ln y_2)^2} \right\}.$$

(In Section 4.3, *-distance was introduced and interpreted graphically.)

Within *-geometry, which is an arithmetic model for Euclidean geometry, one can define counterparts of all Euclidean notions, some of which will be discussed in this section.

A *-translation is a mapping of each bipositive point (x,y) to the bipositive point $(x \overset{\bullet}{+} p, y \overset{\bullet}{+} q)$, that is (px, qy), where p and q are positive constants.[1] It turns out that *-distance is *-invariant; that is, the *-distance between bipositive points is unchanged under all *-translations.

Although $\overset{*}{d}$ is not a metric in the usual sense, it is a *-metric in the sense indicated by the following result.

Theorem

For all bipositive points P_1, P_2, and P_3,
$$\overset{*}{d}(P_1, P_2) \geq \overset{\bullet}{0},$$

$$\overset{*}{d}(P_1,P_2) = \overset{.}{0} \text{ if and only if } P_1 = P_2,$$
$$\overset{*}{d}(P_1,P_2) = \overset{*}{d}(P_2,P_1),$$
$$\overset{*}{d}(P_1,P_2) \overset{.}{+} \overset{*}{d}(P_2,P_3) \geq \overset{*}{d}(P_1,P_3).$$

The bipositive points P_1, P_2, and P_3 are *-collinear provided that at least one of the following holds:

$$\overset{*}{d}(P_2,P_1) \overset{.}{+} \overset{*}{d}(P_1,P_3) = \overset{*}{d}(P_2,P_3),$$
$$\overset{*}{d}(P_1,P_2) \overset{.}{+} \overset{*}{d}(P_2,P_3) = \overset{*}{d}(P_1,P_3),$$
$$\overset{*}{d}(P_1,P_3) \overset{.}{+} \overset{*}{d}(P_3,P_2) = \overset{*}{d}(P_1,P_2).$$

If $x_1 = x_2 = x_3$ or $y_1 = y_2 = y_3$, then P_1, P_2, and P_3 are *-collinear, but other circumstances also engender *-collinearity, as will be seen shortly.

A *-line is a set L of at least two distinct bipositive points such that for any distinct bipositive points P_1 and P_2 in L, a bipositive point P_3 is in L if and only if P_1, P_2, and P_3 are *-collinear.

Theorem

The class of nonvertical *-lines is identical with the class of power functions.[2]

In view of the preceding theorem, one may say that *-geometry is a nonlinear model of Euclidean geometry. (Infinitely-many nonlinear models of Euclidean geometry are constructed in Chapter 10 of [2].)

Two *-lines are *-parallel provided they are identical or have no common point. For example, the *-lines with equations $y = x^2$ and $y = 2x^2$, where $x > 0$, are *-parallel.

Theorem

Two nonvertical *-lines are *-parallel if and only
if they have the same *-slope.

Of course, a *-line that is *-parallel to the horizontal
axis (on *-paper) has *-slope equal to $\overset{\circ}{0}$.

A *-line L_1 is *-perpendicular to a *-line L_2 provided
they have a unique common bipositive point P, and for any bipositive points P_1 on L_1 and P_2 on L_2, distinct from P,
$$\overset{*}{d}(P_1,P) < \overset{*}{d}(P_1,P_2).$$

Theorem

Two nonvertical *-lines are *-perpendicular if and
only if the geometric product of their *-slopes is
$\overset{*}{-1}$.

Theorem

If two *-perpendicular *-lines L_1 and L_2 intersect
at bipositive point P, then for all bipositive
points P_1 on L_1 and P_2 on L_2,
$$[\overset{*}{d}(P_1,P)]^{\overset{\cdot}{2}} \overset{\cdot}{+} [\overset{*}{d}(P,P_2)]^{\overset{\cdot}{2}} = [\overset{*}{d}(P_1,P_2)]^{\overset{\cdot}{2}}.$$

The *-arc-length of a suitably behaved arc can be defined
with the usual limit process by partitioning the arc and using
the geometric sum of the *-distances between successive bipositive points. If a bipositive function f has a continuous *-
derivative on a positive interval [r,s], then the *-arc-length
of f on [r,s] turns out to be
$$\overset{*}{\int}_r^s \sqrt{\overset{\cdot}{1} \overset{\cdot}{+} (\overset{*}{D}f)^{\overset{\cdot}{2}}}.$$

As expected, the *-distance between two distinct bipositive points P_1 and P_2 is equal to the *-arc-length of the *-line segment connecting P_1 and P_2.

Using *-distance and geometric arithmetic, one can define the *-conics, the study of which may be facilitated by the *-vectors defined in the next chapter.

Though the Cartesian and *-geometries are structurally equivalent, it is just as profitable to distinguish them as it is to distinguish the classical and *-calculi.

N O T E S

1. Clearly, applying a *-translation to the points of a bipositive function is equivalent to making a change of scales (or units) in the function.

2. The definition of "vertical" given in Section 6.2 applies here also.

CHAPTER 7

Bigeometric Vectors and Centroids

7.1 BIGEOMETRIC VECTORS

The set of all classical translations can be made into a
vector space in the following way. For any numbers a and b,
let the classical translation that maps each point (x,y) to
(x + a, y + b) be denoted by v[a;b] and be called a <u>classical
vector</u>. Define the <u>sum</u> of v[a;b] and v[c;d] to be v[a + c;
b + d], and define the <u>scalar product</u> of a number k and v[a;b]
to be v[ka; kb]. Then this system is a vector space over the
field of classical arithmetic. One may also define the <u>norm</u>
of v[a;b] to be the number

$$\sqrt{a^2 + b^2} \, ,$$

which equals the classical distance between (a,b) and (0,0).

Every classical vector v[a;b] is <u>rectilinear</u> in the sense
that for any point (x,y), the following points are collinear:
(x,y), its image (x + a, y + b), and the latter's image
(x + a + a, y + b + b).

The set of all *-translations can also be made into a
vector space. For any positive numbers p and q, let the *-
translation that maps each bipositive point (x,y) to (x $\dotplus$ p,
y $\dotplus$ q) be denoted by $\overset{*}{v}$[p;q] and be called a *-vector. Define
the <u>sum</u> of $\overset{*}{v}$[p;q] and $\overset{*}{v}$[s;t] to be $\overset{*}{v}$[p $\dotplus$ s; q $\dotplus$ t], and define
the <u>scalar product</u> of a positive number k and $\overset{*}{v}$[p;q] to be

$\overset{*}{v}[k \overset{\cdot}{\times} p; k \overset{\cdot}{\times} q]$. Then this system is a vector space over the field of geometric arithmetic. One may also define the <u>norm</u> of $\overset{*}{v}[p;q]$ to be the number

$$\overset{\cdot}{\sqrt{p^{\overset{\cdot}{2}} \overset{\cdot}{+} q^{\overset{\cdot}{2}}}},$$

which equals the *-distance between (p,q) and $(\overset{\cdot}{0},\overset{\cdot}{0})$.

Each *-vector $\overset{*}{v}[p;q]$ is <u>*-rectilinear</u> in the sense that for any bipositive point (x,y), the bipositive points (x,y), $(x \overset{\cdot}{+} p, y \overset{\cdot}{+} q)$, and $(x \overset{\cdot}{+} p \overset{\cdot}{+} p, y \overset{\cdot}{+} q \overset{\cdot}{+} q)$ are *-collinear. If $p = \overset{\cdot}{0}$ or $q = \overset{\cdot}{0}$ or $p = q$, then $\overset{*}{v}[p;q]$ is <u>rectilinear</u> because for any bipositive point (x,y), the bipositive points (x,y), $(x \overset{\cdot}{+} p, y \overset{\cdot}{+} q)$, and $(x \overset{\cdot}{+} p \overset{\cdot}{+} p, y \overset{\cdot}{+} q \overset{\cdot}{+} q)$ are collinear. However, if $p \neq \overset{\cdot}{0}$, $q \neq \overset{\cdot}{0}$, and $p \neq q$, then $\overset{*}{v}[p;q]$ is <u>curvilin-ear</u> because it is not rectilinear. Thus, most *-vectors are curvilinear.

7.2 BIGEOMETRIC CENTROIDS

In a document sent to Eratosthenes, Archimedes explained in detail how he used classical centroids to discover deep geometric theorems.[1] Believing that centroids will one day a-gain prove to have considerable heuristic value in pure mathe-matics, we devote this section to a novel definition of the classical centroid and to a definition of the new *-centroid. Both centroids will be used in Chapter 8 on least-squares me-thods.

Since the classical distance from a point $P(x,y)$ to a point $P_1(x_1,y_1)$ is

$$\sqrt{(x - x_1)^2 + (y - y_1)^2}\ ,$$

it is not unreasonable to define the _classical distance_ from a point $P(x,y)$ to n points $P_1(x_1,y_1),\ldots,P_n(x_n,y_n)$ to be

$$\sqrt{[(x - x_1)^2 + (y - y_1)^2] + \cdots + [(x - x_n)^2 + (y - y_n)^2]}\ ,$$

or simply

$$\sqrt{[d(P,P_1)]^2 + \cdots + [d(P,P_n)]^2}\ .$$

For $n = 1$, the preceding expression reduces to $d(P,P_1)$, as it should.

The idea for the following theorem was suggested by an article published by Legendre in 1805.[2]

Theorem

For any n points $P_1,\ldots,P_n$, there is a unique point C that is classically closer than every other point to $P_1,\ldots,P_n$, and the coordinates of C are $(x_1 + \cdots + x_n)/n$ and $(y_1 + \cdots + y_n)/n$.

The point C in the preceding theorem is called the _classical centroid_ of the points $P_1,\ldots,P_n$, and has the following important property:

(1) For each linear function ℓ containing C,

$$[\ell(x_1) - y_1] + \cdots + [\ell(x_n) - y_n] = 0.$$

It is an interesting fact, which was known in substance by Archimedes, that the classical centroid of any given n points of a linear function is contained in that function.

If C is the classical centroid of $P_1, \ldots, P_n$, and $\bar{C}$, $\bar{P}_1$, ..., $\bar{P}_n$ are their images under any given classical translation, then $\bar{C}$ is the classical centroid of $\bar{P}_1, \ldots, \bar{P}_n$.

Our development of the *-centroid is similar to that of the classical centroid.

The *-distance from a bipositive point $P(x,y)$ to n bipositive points $P_1(x_1, y_1), \ldots, P_n(x_n, y_n)$ is defined to be

$$\sqrt[*]{[\overset{*}{d}(P,P_1)]^{\overset{*}{2}} \overset{*}{+} \cdots \overset{*}{+} [\overset{*}{d}(P,P_n)]^{\overset{*}{2}}} \; ,$$

which reduces to $\overset{*}{d}(P,P_1)$ for n = 1.

Theorem

> For any bipositive points $P_1, \ldots, P_n$, there is a unique bipositive point C that is *-closer than every other bipositive point to $P_1, \ldots, P_n$, and the coordinates of C are $(x_1 \overset{*}{+} \cdots \overset{*}{+} x_n) \overset{*}{/} n$ and $(y_1 \overset{*}{+} \cdots \overset{*}{+} y_n) \overset{*}{/} n$. (Of course, the first coordinate of C is the geometric average of $x_1, \ldots, x_n$ and the second coordinate is the geometric average of $y_1, \ldots, y_n$.)

The bipositive point C in the preceding theorem is the *-centroid of the bipositive points $P_1, \ldots, P_n$, and has the following property:

(2) For each power function p containing C,
$$[p(x_1) \overset{*}{-} y_1] \overset{*}{+} \cdots \overset{*}{+} [p(x_n) \overset{*}{-} y_n] = \overset{*}{0}.$$

The *-centroid of any given n bipositive points of a power function is contained in that function.

If C is the *-centroid of n bipositive points $P_1, \ldots, P_n$, and $\overset{*}{C}$, $\overset{*}{P}_1$, ..., $\overset{*}{P}_n$ are their images under any given *-translation, then $\overset{\star}{C}$ is the $\star$-centroid of $\overset{\star}{P}_1, \ldots, \overset{\star}{P}_n$.

N O T E S

1. A palimpsest of Archimedes' document, commonly called *The Method*, was discovered in 1906 by J. L. Heiberg.

2. Legendre's article is entitled "Sur la Méthode des moindres quarrés," a translation of which appears in *A Source Book in Mathematics*, Vol. II, ed. David E. Smith (Dover reprint, 1959), pp. 576-9.

CHAPTER 8

The Bigeometric Method of Least Squares

8.1 INTRODUCTION

> [The classical method of least squares is] "one of
> the most fundamental cornerstones of statistical
> methods."
>
> Cornelius Lanczoz[1]

Ever since the classical method of least squares was de-

veloped by Gauss, Laplace, Legendre, and others, there have

been disagreements over the method of measuring the deviation

between an estimate and the "true" value. Laplace claimed

that the absolute value of the difference should be used.

Gauss maintained that the square of the difference is a better

choice and he made the following remark in 1820:

> "If you object that this is arbitrary, we readily a-
> gree. The question with which we are concerned is
> vague in its very nature; it can only be made pre-
> cise by pretty arbitrary principles. Determining a
> magnitude by observation can justly be compared to
> a game in which there is a danger of loss but no
> hope of gain.... But if we do so, to what can we
> compare an error which has actually been made? That
> question is not clear, and its answer depends in
> part on convention. Evidently the loss in the game
> can't be compared directly to the error which has
> been committed, for then a positive error would rep-
> resent a loss, and a negative error a gain. The
> magnitude of the loss must on the contrary be evalu-
> ated by a function of the [difference], whose value
> is always positive. Among the infinite number of
> functions satisfying these conditions, it seems nat-
> ural to choose the simplest, which is, beyond con-
> tradiction, the square of the [difference]."[2]

Despite the many sophisticated attempts to justify Gauss'

choice by various statistical-probabilistic techniques, some mathematicians still contend that the least-squares criterion is purely arbitrary. Although we are not concerned here with this controversy, we shall attempt to present the classical method of least squares in a natural manner. And then we shall use geometric arithmetic to explain the bigeometric method of least squares, which provides a rationale for a technique well-known to scientists.

N O T E S

1. Cornelius Lanczoz, *Albert Einstein and the Cosmic World Order* (Interscience, 1965), p. 65.

2. Gauss' remark is quoted in Ian Hacking's book, *Logic of Statistical Inference* (Cambridge University Press, 1965), p. 175.

8.2 THE CLASSICAL METHOD OF LEAST SQUARES

Let f be a discrete function with arguments $a_1, \ldots, a_n$, and let h be any function whose arguments include all the a_i. Since h is to be conceived as an "estimate" for f, our first problem is to settle on a suitable method of defining the "distance" or "deviation" between h and f in terms of their values at the a_i. Therefore we turn our attention to the sequences

$$h(a_1), \ldots, h(a_n)$$

and

$$f(a_1), \ldots, f(a_n).$$

Each of those two sequences of n numbers may be conceived as a point in n-dimensional Euclidean space, in which the distance between the two points is given by

$$\sqrt{[h(a_1) - f(a_1)]^2 + \cdots + [h(a_n) - f(a_n)]^2} \; .$$

Accordingly we choose that number to represent the <u>classical distance</u> between the given functions h and f. For n = 1, the preceding expression reduces to the classical distance $|h(a_1) - f(a_1)|$ between the points $(a_1, h(a_1))$ and $(a_1, f(a_1))$.

Now we are prepared to state the theorem that will motivate an important definition.

<u>Theorem</u>

> For each discrete function f with at least two distinct points, there is a unique linear function ℓ that is classically closer than every other linear function to f.

The linear function ℓ in the foregoing theorem is called the linear function that best-fits f by the classical method of least squares, henceforth abbreviated as <u>the linear function that best-fits f classically</u>; and ℓ has two important properties, which, however, do not characterize it:[1]

(1) $[\ell(a_1) - f(a_1)] + \cdots + [\ell(a_n) - f(a_n)] = 0$.

(2) ℓ contains the classical centroid of f.

Thus, the linear function that best-fits f classically is the linear function classically closest to f; and that linear function contains the classical centroid of f, which is the point classically closest to f.

The classical slope of the linear function that best-fits f classically is often used to indicate the "overall linear direction" of f.

The foregoing presentation is a natural extension of the idea that motivated our definition of the classical centroid in Section 7.2.

Now let us consider the problem of classically fitting nonlinear functions to a discrete function f: $\{(a_1,b_1),\ldots,(a_n,b_n)\}$. Consider any set S of functions whose arguments include all the a_i. If there exists a unique member h of S that is classically closer than every other member of S to f, then h is called, for brevity, <u>the member of S that best-fits f classically</u>. (Please note that the members of S are not assumed to be linear or discrete.)

For example, scientists often wish to fit a power function to f. (Here S is the set of all power functions, and we assume that f is bipositive and has at least two distinct points.) Although the classical method of least squares may be used to solve this problem, the calculations are usually quite messy. Accordingly, scientists often adopt the following seemingly makeshift technique.

First f is transformed to $\{(\ln a_1, \ln b_1), \ldots, (\ln a_n, \ln b_n)\}$, denoted by $\bar{f}$. Similarly, each power function p is transformed to a linear function by transforming each point (x,y) in p to $(\ln x, \ln y)$. Then, the linear function ℓ that best-fits $\bar{f}$ classically is found. And finally, ℓ is trans-

formed to a power function by transforming each point (u,v) in
ℓ to (exp u, exp v). In Section 8.4, we shall return to this
technique.

N O T E

1. The reader may wish to compare item (1) with the items labeled (1) in Sections 3.3 and 7.2.

8.3 THE BIGEOMETRIC METHOD OF LEAST SQUARES

Let f be a discrete bipositive function with arguments
$a_1, \ldots, a_n$.

If h is any bipositive function whose arguments include
all the a_i, then the *-distance between h and f is defined to
be

$$\sqrt{[h(a_1) \overset{\cdot}{-} f(a_1)]^{\overset{\cdot}{2}} \overset{\cdot}{+} \cdots \overset{\cdot}{+} [h(a_n) \overset{\cdot}{-} f(a_n)]^{\overset{\cdot}{2}}}\ .$$

For n = 1, the preceding expression reduces to the *-distance
$\overset{\cdot}{|}h(a_1) \overset{\cdot}{-} f(a_1)\overset{\cdot}{|}$ between $(a_1,\ h(a_1))$ and $(a_1,\ f(a_1))$.

Theorem

> For each discrete bipositive function f with at
> least two distinct points, there is a unique power
> function p that is *-closer than every other power
> function to f.

The power function p in the foregoing theorem is called
the power function that best-fits f by the bigeometric method
of least squares, abbreviated as the power function that best-

<u>fits f bigeometrically</u>; and p has two important properties, which, however, do not characterize it:[1]

    (1) $[p(a_1) \overset{\text{-}}{\text{-}} f(a_1)] \overset{\cdot}{+} \cdots \overset{\cdot}{+} [p(a_n) \overset{\text{-}}{\text{-}} f(a_n)] = \overset{\circ}{0}.$

    (2) p contains the *-centroid of f.

Thus, the power function that best-fits f bigeometrically is the power function *-closest to f; and that power function contains the *-centroid of f, which is the bipositive point *-closest to f.

The *-slope of the power function that best-fits f bigeometrically may be useful for indicating the "overall *-direction" of f.

Now let S be any set of bipositive functions whose arguments include all the a_i. If there exists a unique member h of S that is *-closer than every other member of S to f, then h will be called, for brevity, <u>the member of S that best-fits</u> <u>f bigeometrically</u>. (Please note that the members of S are not assumed to be discrete or to be power functions.)

The bigeometric method of least squares is a natural extension of the idea that motivated our definition of the *-centroid in Section 7.2.

N O T E

1. The reader may wish to compare item (1) with the items labeled (3) in Section 3.3 and (2) in Section 7.2.

8.4 THE RELATIONSHIP BETWEEN THE TWO METHODS

Let f be a discrete bipositive function, and let S be any set of bipositive functions whose arguments include all the arguments of f. For each bipositive function g, let $\bar{g}$ be the function consisting of all points (ln x, ln y) for which (x,y) is in g. Let $\bar{S}$ be the set of all functions $\bar{g}$ for which g is in S. Let $\overset{*}{L}(S;f)$ be the member of S that best-fits f bigeometrically, if it exists; and let $L(\bar{S};\bar{f})$ be the member of $\bar{S}$ that best-fits $\bar{f}$ classically, if it exists.

Then $\overset{*}{L}(S;f)$ and $L(\bar{S};\bar{f})$ coexist, and if they do exist,

$$\overline{\overset{*}{L}(S;f)} = L(\bar{S};\bar{f}).$$

It follows that if $\overset{*}{L}(S;f)$ exists, it can be found by using the classical method of least squares to select the function in $\bar{S}$ that best-fits $\bar{f}$ and then transforming that function by transforming each of its points (u,v) to (exp u, exp v).

Thus, the seemingly makeshift technique described in Section 8.2 is but the bigeometric method of least squares for fitting a power function to a given discrete bipositive function, a fact that amazed us when we first noticed it.

CHAPTER 9

Related Matters

9.1 INTRODUCTION

Among other things, this last chapter contains a brief discussion of *-complex-numbers. The reader might enjoy investigating other topics related to the *-calculus; for example, *-differential equations, Taylor series in the context of *-calculus, *-calculus of functions of several real variables, the theory of functions of a *-complex variable, or the application of *-calculus to science and engineering.

9.2 BIGEOMETRIC COMPLEX-NUMBERS

The *-complex-number system consists of all bipositive points, for which the following addition and multiplication operations are defined:

$$(a_1,b_1) \overset{*}{+} (a_2,b_2) = (a_1 \overset{\cdot}{+} a_2,\ b_1 \overset{\cdot}{+} b_2),$$

$$(a_1,b_1) \overset{*}{\times} (a_2,b_2) =$$
$$((a_1 \overset{\cdot}{\times} a_2) \overset{\cdot}{-} (b_1 \overset{\cdot}{\times} b_2),\ (a_1 \overset{\cdot}{\times} b_2) \overset{\cdot}{+} (b_1 \overset{\cdot}{\times} a_2)).$$

It is interesting to observe that

$$(a_1,b_1) \overset{*}{+} (a_2,b_2) = (a_1 \cdot a_2,\ b_1 \cdot b_2),$$

$$(a_1,b_1) \overset{*}{\times} (a_2,b_2) =$$

$$((a_1^{\ln a_2})/(b_1^{\ln b_2}),\ (a_1^{\ln b_2}) \cdot (b_1^{\ln a_2})).$$

Because the system is a field, one can define subtraction and division operations that are "inverses" of $\overset{*}{+}$ and $\overset{*}{\times}$, respectively.

Since

$$(\dot{0},\dot{1}) \overset{*}{\times} (\dot{0},\dot{1}) = (\overset{\cdot}{\underset{\cdot}{\pm}}\dot{1},\dot{0}),$$

we define $\overset{*}{i}$ to be $(\dot{0},\dot{1})$.

The *-modulus of a bipositive point (a,b) is defined to be the *-distance between (a,b) and $(\dot{0},\dot{0})$, which equals

$$\overset{\cdot}{\sqrt{a^{\overset{\cdot}{2}} \overset{\cdot}{+} b^{\overset{\cdot}{2}}}} .$$

9.3 WEIGHTED CALCULI AND META-CALCULI

In this section we shall briefly discuss the weighted calculi and meta-calculi that were first presented in [5] and [6].

By a classical weight function we mean any continuous positive-valued function on R. There are, of course, infinitely-many classical weight functions.

In [5], it is shown that each classical weight function ω can be used to construct a system of calculus, called a weighted classical calculus, whose operators are the classical calculus operators "weighted" by ω. In each weighted classical calculus, the natural average is a weighted arithmetic average, the integral is a Stieltjes integral, and the derivative is a classical derivative of one function with respect to another.

Furthermore, in [6], it is shown that each ordered pair of classical weight functions can be used to construct a system of calculus, called a meta-calculus, that transcends the classical calculus, for example in the following manner. In each meta-calculus the gradient of a function f on an interval [r,s] depends on *all* the points (x,f(x)) for which r ≤ x ≤ s, whereas the classical gradient [f(s) - f(r)]/(s - r) depends only on the endpoints (r,f(r)) and (s,f(s)).[1]

By a *-weight function we mean any continuous function on R_+ whose values are all greater than Ȯ. There are infinitely-many *-weight functions, each of which is bipositive.

Each *-weight function w can be used to construct a system of calculus, called a weighted *-calculus, whose operators are the *-calculus operators "weighted" by w.

Indeed, each non-Newtonian calculus can be "weighted" in a manner explained fully in [5]. It turns out that the well-known weighted geometric average, weighted harmonic average, and weighted power averages fit naturally into certain weighted non-Newtonian calculi.

N O T E

1. The meta-calculi arose from the problem of measuring stock-price performance when taking all intermediate prices into account.

9.4 AN INSIGHT BY BOSCOVICH

J. F. Scott of St. Mary's College, London, has observed that the following astonishing quotation indicates Roger Joseph Boscovich (1711 - 1787) was capable of grasping the possibility of a non-Euclidean geometry.

> "But if some mind very different from ours were to look upon some property of some curved line as we do on the evenness of a straight line, he would not recognize as such the evenness of a straight line; nor would he arrange the elements of his geometry according to that very different system, and would investigate quite other relationships as I have suggested in my notes.
>
> "We fashion our geometry on the properties of a straight line because that seems to us to be the simplest of all. But really all lines that are continuous and of a uniform nature are just as simple as one another. Another kind of mind which might form an equally clear mental perception of some property of any one of these curves, as we do of the congruence of a straight line, might believe these curves to be the simplest of all, and from that property of these curves build up the elements of a very different geometry, referring all other curves to that one, just as we compare them to a straight line. Indeed, these minds, if they noticed and formed an extremely clear perception of some property of, say, the parabola, would not seek, as our geometers do, to *rectify* the parabola, they would endeavour, if one may coin the expression, to *parabolify* the straight line."

This quotation indicates to us that Boscovich was capable of grasping the possibility of a non-Newtonian calculus.[1]

N O T E

1. In April of 1973, Robert Katz and I encountered the Boscovich quotation and Scott's remark in his article "Boscovich's Mathematics," which appears in Lancelot Law Whyte's anthology *Roger Joseph Boscovich* (George Allen & Unwin Ltd., and Fordham University Press, 1961). Scott apparently obtained the Boscovich quotation from H. V. Gill's book, *Roger Joseph Boscovich:*

Forerunner of Modern Physical Theories (Dublin, 1941), pp. 50-53.

Perhaps this disclosure of Boscovich's insight will inspire some historian to investigate the matter fully, especially since Boscovich refers to his "notes" on the subject. The article on Boscovich in the magnificent *Dictionary of Scientific Biography*, edited by Charles Coulston Gillispie, sheds little light on the issue at hand.

9.5 CONCLUSION

> "There seem to be two kinds of discovery. In one kind, the goal is given first and then the mind goes from the goal to the means, that is, from the question to the solution. In the other kind, the mind goes from the means to the goal, that is, the mind first discovers a fact and then seeks a use for it. In mathematics, and elsewhere, most significant discoveries are of the second kind. As Hadamard has put it, 'Practical application is found by not looking for it, and one can say that the whole progress of civilization rests on that principle.' An outstanding example in mathematics is the exhaustive study of the conics by the Greeks, and then, some two thousand years later, Kepler's stunning application of the Greek findings to the movement of the planets in the solar system. The physicist and artist Duhem once compared Hadamard to a landscape painter who in his studio creates a landscape painting and then leaves the studio to find in nature some landscape fitting his painting."
>
> Howard Eves[1]

At the Seminar on Bayesian Inference in Econometrics (sponsored by the National Bureau of Economic Research and the National Science Foundation) in May of 1980, Professor James R. Meginniss, then of Harvey Mudd College and the Claremont Graduate School, showed how non-Newtonian calculus can be applied to subjective probability, utility, and Bayesian analysis. (See [4].)

Of course, we can only speculate as to future applications of the non-Newtonian calculi. Perhaps they can be used to define new scientific concepts, to yield new or simpler scientific laws, to solve heretofore unsolved problems, or to formulate and solve new problems.

In the three centuries since the creation of the classical calculus, the greatest mathematician was Gauss, with whose wise words we conclude this book.

> "In general the position as regards all such new calculi is this - That one cannot accomplish by them anything that could not be accomplished without them. However, the advantage is, that, provided such a calculus corresponds to the inmost nature of frequent needs, anyone who masters it thoroughly is able - without the unconscious inspiration of genius which no one can command - to solve the respective problems, yea, to solve them mechanically in complicated cases in which, without such aid, even genius becomes powerless. Such is the case with the invention of general algebra, with the differential calculus, and in a more limited region with Lagrange's calculus of variations, with my calculus of congruences, and with Möbius's calculus. Such conceptions unite, as it were, into an organic whole countless problems which otherwise would remain isolated and require for their separate solution more or less application of inventive genius."[2]

N O T E S

1. This remark appears on pages 167, 168 of Howard Eves' entertaining book, *Mathematical Circles Squared* (Prindle, Weber, and Schmidt, 1972).

2. Gauss' remark appears in Robert E. Moritz's book, *On Mathematics and Mathematicians* (Dover reprint, 1958), pp. 197-8. The Möbius calculus referred to by Gauss is the so-called barycentric calculus, which is geometry, not calculus in the usual sense.

BIBLIOGRAPHY

1. Katz, R. *Axiomatic Analysis*. Rockport, MA: Mathco, 1964.
This textbook, which was prepared under the general editorship
of Professor David V. Widder, contains an original approach to
basic logic and a novel axiomatic treatment of the real number
system.

2. Grossman, M. and Katz, R. *Non-Newtonian Calculus*. Rockport, MA: Mathco, 1972.
Included in this book, which was the first publication on non-
Newtonian calculus, are discussions of nine specific non-Newtonian calculi, the general theory of non-Newtonian calculus,
and heuristic guides for the application thereof.

3. Grossman, M. *The First Nonlinear System of Differential
and Integral Calculus*. Rockport, MA: Mathco, 1979.
This book contains a detailed account of the geometric calculus, which was the first of the non-Newtonian calculi. Also
included are discussions of the analogy that led to the discovery of that calculus, and some heuristic guides for its application.

4. Meginniss, J. R. "Non-Newtonian Calculus Applied to Probability, Utility, and Bayesian Analysis." *Proceedings of the
American Statistical Association*: Business and Economics Statistics Section (1980), pp. 405-410.
This paper presents a new theory of probability suitable for
the analysis of human behavior and decision making. The theory is based on the idea that subjective probability is govern-
ed by the laws of a non-Newtonian calculus and one of its corresponding arithmetics.

5. Grossman, J., Grossman, M., and Katz, R. *The First Systems of Weighted Differential and Integral Calculus*. Rockport, MA: Archimedes Foundation, 1980.
This monograph reveals how weighted averages, Stieltjes integrals, and derivatives of one function with respect to another
can be linked to form systems of calculus, which are called
weighted calculi because in each such system a weight function
plays a central role.

6. Grossman, J. *Meta-Calculus: Differential and Integral*.
Rockport, MA: Archimedes Foundation, 1981.
This monograph contains a development of systems of calculus,
called meta-calculi, that transcend the classical calculus,
for example in the following manner. In each meta-calculus
the gradient, or average rate of change, of a function f on an
interval [r,s] depends on *all* the points (x,f(x)) for which
$r \leq x \leq s$, whereas the classical gradient $[f(s) - f(r)]/(s - r)$
depends only on the endpoints (r,f(r)) and (s,f(s)). The meta-
calculi arose from the problem of measuring stock-price performance when taking all intermediate prices into account.

96

7. Grossman, J., Grossman, M., and Katz, R. *Averages: A New Approach*. Rockport, MA: Archimedes Foundation, 1983.
This monograph is primarily concerned with a comprehensive family of averages (unweighted and weighted) of functions that arose naturally in the development of non-Newtonian calculus and weighted non-Newtonian calculus. The monograph also contains discussions of some heuristic guides for the appropriate use of averages and an interesting family of means of two positive numbers.

LIST OF SYMBOLS

I N D E X

www.ingramcontent.com/pod-product-compliance
Lightning Source LLC
Chambersburg PA
CBHW022112210326
41521CB00028B/316